THE EARLY BRITISH TIN INDUSTRY

Sandy Gerrard

THE EARLY BRITISH
TIN INDUSTRY

Sandy Gerrard

TEMPUS

First published 2000

PUBLISHED IN THE UNITED KINGDOM BY:

Tempus Publishing Ltd
The Mill, Brimscombe Port
Stroud, Gloucestershire GL5 2QG

PUBLISHED IN THE UNITED STATES OF AMERICA BY:

Tempus Publishing Inc.
2A Cumberland Street
Charleston, SC 29401

Tempus books are available in France, Germany and Belgium
from the following addresses:

Tempus Publishing Group
21 Avenue de la République
37300 Joué-lès-Tours
FRANCE

Tempus Publishing Group
Gustav-Adolf-Straße 3
99084 Erfurt
GERMANY

Tempus Publishing Group
Place de L'Alma 4/5
1200 Brussels
BELGIUM

© Sandy Gerrard 2000

British Library Cataloguing in Publication Data.
A catalogue record for this book is available from the British Library.

ISBN 0 7524 1452 6

Typesetting and origination by Tempus Publishing.
PRINTED AND BOUND IN GREAT BRITAIN.

Contents

Acknowledgements 6

Illustrations 7

1 Introduction 10

2 Tinworking from prehistory until 1066 14

3 The medieval and early modern period 25

4 Tinworks and tinbounds 40

5 Streamworks 60

6 Mining 81

7 Stamping, crushing and dressing 104

8 Smelting 129

9 Sites to visit 140

Glossary 149

Further reading 151

Index 154

Acknowledgements

During the past 20 years I have come into contact with a large number of people who have helped me in a variety of ways to first appreciate, then wonder at and finally understand the complexity of the archaeological evidence relating to the tin industry. First amongst these must be David Austin who inadvertently introduced me to the subject when I must have been at an impressionable age. Since that time I have been encouraged, inspired and helped by a variety of friends, colleagues and acquaintances.

Amongst these are Alex Black, Mary Braid, Kath Brewer, Tony Brown, Jeremy Butler, David Cranstone, Ann Dick, Bryan Earl, Ted Fitch, Tom Greeves, Debbie Griffiths, John Gurdler, Peter Herring, Dave Hooley, the Late Andrew Lawson, Jacky Nowakowski, Cathy O'Mahoney, Clue and Jane Masters, Oliver Padel, Roger Penhallurick, Derrick Popple, Ann Preston-Jones, Athos Pittordou, Peter Rose, Adam Sharpe and John Smith.

In the course of my fieldwork I have met many landowners who have always been most supportive and have generously allowed me access to their property. In this respect I would especially like to thank: the Late Mr Clarke and Mrs Clarke, Mr and Mrs Davey, David Hyde, Mr and Mrs Lovelock, Mr Rowe, Mr Saunders, Mr and Mrs Wherry, Mr and Mrs Wills, Mr Colin Sturmer (Duchy of Cornwall) Michael Green (South West Water Services Ltd) and Chris Marrow (Forest Enterprise). Without their support much of the archaeological evidence upon which this book relies so heavily could never have been researched.

The staff of the County Record Offices, Royal Institution of Cornwall, Redruth Local Studies Library and the Public Record Office immensely facilitated the documentary research.

A smaller number of colleagues have been directly involved with the production of this book and in this respect I would like to thank: Nick Johnson and Steve Hartgroves for providing ready access to the Cornwall Sites and Monuments Register and generously allowing me to use some of their very fine aerial photographs; Frances Griffith (Devon County Council) for her photograph of Vitifer and extensive use of the County Sites and Monuments Register; Dennis Lethbridge who provided photographs and other information; Phil Newman and Chris Powell who have both generously allowed me to use their reconstruction illustrations.

I would also like to thank Peter Kemmis Betty (Tempus), who asked me write this book and who, together with Anne Phipps, thereafter helped bring it to life.

Finally, a considerable debt is owed to my wife Helen who as well as helping with much of the fieldwork has also checked through the text and provided inspiration throughout the process of researching and writing this work.

Illustrations

Text figures

1   Location of the stannaries and stannary towns
2   The variety of field evidence within the tin production framework
3   A human skeleton found at Perran-ar-Worthal
4   Map of Dartmoor and Foweymore showing the location of tin streamworks and prehistoric settlements
5   A bowl furnace
6   The tin pan from the streamwork at Treloy
7   Dowsing and the digging of prospecting pits
8   Prospecting pits near an eluvial streamwork west of Black Rock
9   Prospecting and lode-back pits south of Black Tor
10   Prospecting features around Hart Tor
11   Alluvial streamwork at Colliford
12   Eluvial streamwork on the northern side of Penkestle Moor
13   Leats and reservoirs on Penkestle Moor
14   Simplified plan showing the character of a selection of tinners' buildings
15   Map of Dartmoor and Foweymore showing the position of tin streamworks and medieval settlements
16   Map highlighting the number of tinworks in each stannary in the sixteenth century and seventeenth centuries
17   Distribution of sixteenth-century tinworks
18   Maps showing the weight of tin produced from different parts of the South-West at three different dates
19   Alluvial streamwork at Minzies Down
20   Alluvial tin streamwork on the Hart Tor Brook
21   Developmental sequence illustrating the manner in which the characteristic parallel dumps found at many alluvial streamworks were formed
22   Alluvial streamwork at Lydford Woods
23   Alluvial streamworks at Stanlake and Hart Tor Brook
24   Section through the alluvial tin streamwork earthworks at Lydford Woods
25   Simplified interpretative plan of an area of alluvial streamwork earthworks on the Hart Tor Brook
26   Retained dumps on the Langcombe Brook streamwork
27   Waste dump buried by later sediment
28   Eluvial streamwork near South Carne

29   Eluvial streamwork on West Moor
30   Eluvial streamwork at Harrowbridge (Western)
31   Eluvial streamwork at Westmoorgate
32   Eluvial streamwork at Harrowbridge (Eastern)
33   Tinwork on Goonzion Down
34   Tinwork at Kit Hill
35   Lode-back tinwork at Black Tor
36   Lode-back tinwork on Morvah Hill
37   Distribution of lode-back tinworks on Dartmoor and Foweymore
38   Lode-back tinworks at South Phoenix
39   Lode-back tinwork and field system at Lanyon
40   Lode-back tinwork at Hobb's Hill
41   Eluvial streamwork and lode-back workings on southern Penkestle Moor
42   Tinworks in the Meavy and Newleycombe Valleys
43   Complex mining landscape at Vitifer
44   Distribution of openworks on Dartmoor and Foweymore
45   West Colliford openwork
46   Excavated trench across West Colliford openwork
47   Openworks and earlier eluvial streamworks at Redhill Downs and Stanlake
48   A simplified view of a tin mine showing the character of underground workings
49   Small shaft and adit mine at Trebinnick
50   Whim platforms at Brown Gelly and Chagford Common
51   Contemporary illustration showing different means of access to underground workings
52   Stamping mill (reconstruction)
53   Wet stamping machinery
54   Stamping mill and associated dressing floor
55   Distribution of stamping mills
56   Simplified plans of a selection stamping mills
57   Distribution of mortar stones on Dartmoor and Foweymore
58   Tin mills at Retallack
59   The openwork, tin mill, leats, shafts and alluvial streamwork at West Colliford
60   Excavations at West Colliford stamping mill
61   Interior of the stamping mill at West Colliford
62   Triangular shaped buddle being used to dress tin
63   Plan of two buddles excavated at West Colliford
64   Slime pond, buddles and mill at West Colliford
65   Stamping mills and streamwork at Black Tor Falls
66   Blowing house (reconstruction)
67   Blowing house and owners' marks
68   Distribution of blowing houses
69   A selection of blowing house plans
70   Mould stone at Upper Yealm blowing house
71   Location of places to visit

72   Two blowing houses in the Yealm Valley
73   Tinworks near Minions
74   Streamworks on West Moor

Colour Plates

1   Prospecting gully cutting through a prehistoric stone row at Hart Tor
2   Triangular tinners' buildings at Leskernick
3   Tinners' building at Beckamore Combe
4   Beehive hut at Lade Hill Brook
5   Tinners' cache at Stonetor Brook
6   Excavated cache at East Colliford
7   Medieval longhouse at Bunning's Park
8   An eluvial streamwork (reconstruction)
9   Alluvial streamwork earthworks in the Plym Valley
10   Eluvial streamwork at Beckamore Combe
11   Waste sands and silts lying between two earlier stone dumps at Lydford Woods
12   Retained dumps at Brim Brook alluvial streamwork
13   Eluvial streamwork at Redhill
14   Retained dumps at the Great Links Tor eluvial streamwork
15   Openworks at Challacombe
16   Mortar stone at Norsworthy Left Bank stamping mill
17   Mortar stone at Black Tor Falls Left Bank stamping mill
18   Mortar stone at Retallack
19   Mortar stones and mould stone at Little Horrabridge
20   Norsworthy Left Bank stamping mill
21   Excavated blowing house at Upper Merrivale
22   Excavated stamping mill at West Colliford
23   Gobbett tin mill
24   Crazing and mortar stones at Retallack
25   Two buddles at West Colliford
26   Mould stone at Retallack
27   Furnace and mould stone at Lower Merrivale
28   The float and mould stones at Lower Merrivale blowing house
29   Stamping mill at Black Tor Falls Left Bank

1 Introduction

Many people who have wandered in the wild upland spaces of the South-West of Britain, must have observed the numerous pits, scars and spoilheaps that attest to centuries of human endeavour in the procurement of the precious mineral, tin. To many, the complex nature of these remains, appears almost incomprehensible, and some even assume they are natural features. This work is intended to redress the balance. It explores and explains the myriad of archaeological features associated with the exploitation and processing of tin in Britain, from its earliest beginnings until 1700. The evidence is derived both from surviving field remains, and documentary sources that also include original illustrations and contemporary written accounts. However, the main focus of this evidence is drawn from the surviving archaeology, with certain economic, social and legal aspects of the industry being utilised only to facilitate ease of understanding in specific areas.

Much has been written on the history of the early tin industry, but nowhere else has the archaeological evidence been brought together in a single place. There are however limitations, because the sheer volume of information has meant that it has been necessary to be selective. This said, the purpose of this work is to present a detailed account of the numerous complicated processes and activities involved in finding, collecting and preparing tin for market. In common with many industrial activities, there are a large number of technical terms that cannot be avoided and these are explained within the text.

The archaeological examination of tin exploitation is very much in its infancy. As work progresses, fresh discoveries are sure to enhance our understanding. In particular, our knowledge of the character of prehistoric tinworking is still rudimentary. The need to ascertain the date of individual tinworks is essential, for although documentary evidence tells us that there was a tinwork in a particular location at a certain time, it is usually impossible to correlate the surviving earthworks with the date of the reference. Despite these limitations, the work being carried out in this field has provided us with a meaningful insight into the character of this important industry, and forms the backbone of this book.

The archaeological evidence which forms the basis of this work is best preserved on Dartmoor, where the structures and earthworks survive in a much better condition than in neighbouring Cornwall where later mining and more intensive agricultural activity have so often obliterated the earlier remains. There are, of course, exceptions and a number of well-preserved Cornish sites provide a helpful glimpse into the nature of the industry in that county. Ironically, much of the contemporary documentation relates to Cornwall and specific detailed references generally pertain to sites which no longer exist, whilst for those sites with the most to tell, field archaeologists have little or no collaborative documentation.

Much of the surviving archaeology in Cornwall lies in the area now known as Bodmin Moor. This name is relatively recent in origin and the area was previously known as Foweymore. The original name is adopted here because the area belonged to the stannary district of Foweymore. In the medieval and later periods Devon and Cornwall were subdivided into nine administrative units called stannaries, each served by an administrative centre called a stannary town (1).

The economies of Cornwall and Devon have, for four-thousand years, been influenced to a varying extent by the metallic minerals that lie beneath their soils. From the Bronze Age until the present, the most significant metallic resource has been tin, although copper, gold, silver, lead and iron were also exploited. The demand for tin has led to the landscape being pitted and scarred in the search for, and extraction of, this important metal. British tin is found only in Cornwall and Devon and this limited distribution required the creation of diverse and complex trading routes to transport the metal to the markets.

The history of the tin industry is very complex, with the character of extraction techniques, capital, and markets varying considerably through time. This work is primarily concerned with the archaeological and documentary evidence for 'early' tin extraction. The use of the term 'early' is adopted to define tinworking before the introduction of new and revolutionary extraction methods, which radically altered the character and scale of the industry. The three fundamental developments that were responsible for this phenomenon each made their appearance in the mining scene in the decades either side of 1700. The first of the radical developments was the introduction of gunpowder, which greatly facilitated rock mining and allowed the ore to be exploited more rapidly, and consequently more cheaply. The second was the introduction of the reverbatory furnace, which instead of charcoal, used the cheaper and more plentiful coal to smelt the ore. The third development, was the introduction of steam engines, which were increasingly used for powering pumps, lifting ore and miners to the surface and

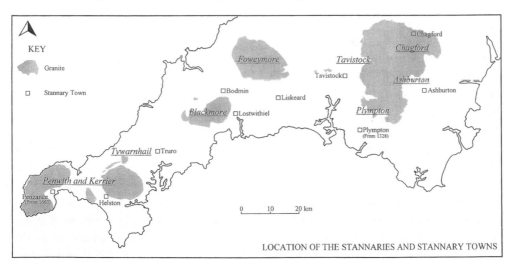

1 Map illustrating the location of the stannaries and stannary towns

driving stamps (ore crushing machinery). These three developments coming within two decades of each other, revolutionised the character of the industry. Output doubled, mines grew steadily deeper, whilst increasing overheads heralded the arrival of large nationally funded companies and corporations, and the end of what may be termed the early tin industry.

Archaeological Research

Compared with most other topics in the South-West of Britain, the study of the early tin industry has until recently been somewhat neglected. During the nineteenth century most of the published work relates to describing artefacts uncovered during streamworking. Notable exceptions to this include an examination of the Lower Yealm blowing house in 1866 by John Kelly; the clearance of the Lower Week Ford house by Robert Burnard, and in Cornwall the work by Bryant at Retallack. It was not until the first half of the twentieth century that R. Hansford Worth carried out a systematic investigation of tin mills. The results of his work were published in a series of articles, but not until the 1970s was the work started by Worth built upon. Over a period of years, Tom Greeves re-examined the industry and considerably enhanced our appreciation of its character and impact on Dartmoor. Until the end of the 1970s research was largely confined to Dartmoor, but this was to change drastically with the building of the new reservoir at Colliford on Foweymore. Where Colliford Lake now lies, a rich archaeological landscape including, streamworks, tin mills, an openwork, shafts and shelters once existed. The archaeological work at Colliford represented the start of a new approach, because here for the first time, mills and shelters were excavated under modern conditions and the tinworks themselves were surveyed in detail. After Colliford, the examination of the tin industry concentrated on recording and understanding the many different types of tinwork and during the 1980s most of this work was carried out in Cornwall. One impact of this change in emphasis was that for the first time tinworks, rather than just the mills, were recognised as being suitable for protection as Scheduled Ancient Monuments.

In the 1990s the focus of research activity moved back to Dartmoor where the methodology developed in Cornwall was adopted and developed. Phil Newman's work on openworks highlighted the urgent need for further research on the moor and in 1991 the Dartmoor Tinworking Research Group was founded to examine all aspects of the industry. To date, a blowing and stamping mill have been excavated and a large number of tinworks surveyed. Areas of tinwork earthworks have also now been examined by the former Royal Commission on the Historical Monuments of England [RCHM(E)] and in the Meavy Valley the author is carrying out a detailed survey that examines the tinworks as part of the overall landscape.

The archaeological remains which together form the evidence for tin extraction and processing are many and varied. Before continuing with detailed explanations of the different components it is useful to illustrate what the elements are and where they fit into the process. This is most easily achieved using a flow chart (2) that illustrates the different components and their place in the process. In subsequent chapters the components of this

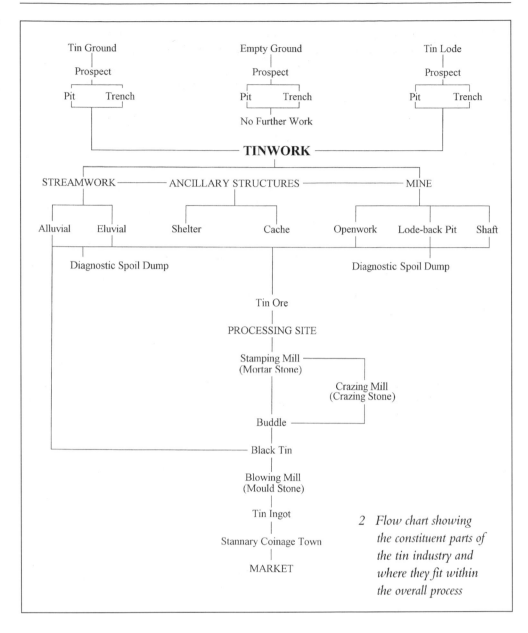

Tin Ground — Prospect — Pit / Trench

Empty Ground — Prospect — Pit / Trench — No Further Work

Tin Lode — Prospect — Pit / Trench

TINWORK

STREAMWORK ———— ANCILLARY STRUCTURES ———— MINE

Alluvial   Eluvial   Shelter   Cache   Openwork   Lode-back Pit   Shaft

Diagnostic Spoil Dump

Diagnostic Spoil Dump

Tin Ore

PROCESSING SITE

Stamping Mill (Mortar Stone)

Crazing Mill (Crazing Stone)

Buddle

Black Tin

Blowing Mill (Mould Stone)

Tin Ingot

Stannary Coinage Town

MARKET

2 Flow chart showing the constituent parts of the tin industry and where they fit within the overall process

chart are examined in detail, but it is important at the outset to understand where in the process they are to be found.

The study of the early tin industry has come a long way in the past 20 years, but because of the manner in which much of the work has been carried out, certain aspects have received more attention than others. This situation has resulted in a bias which is inevitably reflected in this book. Thus, much of the archaeological evidence is derived from work carried out on Foweymore and South-West Dartmoor, whilst much of the documentary evidence is from Cornwall. As research continues and the existing gaps are plugged it is inevitable that the story will be refined.

2 Tinworking from prehistory until 1066

Prehistoric

This chapter examines the archaeological evidence for tin exploitation in Britain in the period up to the Norman Conquest. There is very little documentary material upon which to rely and the picture is largely made up from an accumulation of isolated artefacts which circumstances of more recent years have made available. Much of this evidence comes from central and western Cornwall, although Dartmoor and Foweymore with their rich prehistoric landscapes offer us alternative ways of examining the industry. In broad terms, reworking of many Cornish streamworks in the nineteenth century led to a spate of discoveries, which taken together form a useful if somewhat biased insight into prehistoric tinworking. By contrast, on Dartmoor and to a lesser extent on Foweymore there was no systematic reworking of the earlier streamworks and as a result, the artefacts were not exposed. On a positive note, this probably means that the evidence for early activity still remains on these moors, where as a result, future investigations might prove fruitful. The sources of information include chance finds from streamworks, tin ingots, primitive smelting hearths in prehistoric contexts, environmental evidence and most importantly jewellery and other implements found throughout Britain, which were composed of tin. Later in the period, immediately before the arrival of the Romans in Britain, documentary evidence is available for the first time, and if used carefully provides valuable information.

Considerable analytical work has been carried on the bronze artefacts found throughout Britain and ten distinctive 'industrial phases' of metalworking have been recognised. The details of Bronze-Age metalworking are not being considered here, since the precise character of the finished product did not have any effect on the character of the tinworks. It is sufficient to note that a successful, thriving and developing metalworking tradition in different parts of the country was probably responsible for the exploitation of south-western cassiterite (tin ore) deposits. The British metal market may not have been exclusively served by domestic resources and at least some of the raw material may have come from Brittany and North-West Spain. Northover has noted that the widespread use of scrap metal further confuses the issue of establishing a link between the final product and its source, although Tylecote has suggested that it may be possible to identify artefacts composed of Cornish tin by the distinctive presence of traces of Germanium and Cobalt.

The extensive evidence for tin extraction is briefly considered below and this, combined with the abundantly documented British metalwork tradition suggests that there must have been some form of link between the metal production and manufacturing areas. Detailed electron probe micro-analysis of European bronzes has led Northover to suggest that 'the number of metal sources used at one time was very limited, and there was often only one'. Consequently, trading of metals was on a large scale and this phenomenon he termed the 'metal circulation zone'. For the early Bronze Age, Northover has noted that tin bronzes were probably exclusively produced in Britain from South-Western cassiterite. Whilst in the Middle Bronze Age, the level of tin in many bronzes was radically reduced, possibly as a consequence of 'native hostility', a problem that was overcome by the Later Bronze Age. By the Early Iron Age, the tin output of the South-West was greater than it had been previously. Northover's research was involved with a very small and biased sample of available bronzes. Nevertheless, it indicated the importance of traded metal, and although the significance of the different ore production areas may have varied through time, it appears that tin was extracted and widely exported. Tin was needed by metalsmiths of Bronze-Age and Iron-Age Europe, and evidence for tinworking and its effect on the landscape and economy supports the hypothesis that the South-West of England was an important source of the mineral.

The metal ores of Britain were first exploited in the early second millennium BC and from about this time it must have become abundantly clear to those with the knowledge, that parts of Cornwall and Devon were particularly rich in tin. The prehistoric tinners probably worked the easily accessible alluvial deposits and perhaps only the richest of lodes. This often repeated premise relating to the character of the prehistoric tinworking has become so established in the literature that it is all too often offered as a fact with no explanation or justification. Lewis, however, did note that 'In all young mining districts, even at the present day, the first operation is the digging of the alluvial deposits' and more significantly he drew attention to the discovery of prehistoric artefacts in the streamworks. Hamilton Jenkin, however, made a useful observation regarding these finds noting that 'the account of their provenance is generally too vague to teach us much'. There are welcome exceptions and at Pentewan, a wooden shaft extending from ten feet below the eighteenth-century land surface to the tin ground was associated with an arrowhead and a small chisel, and at Perran-ar-Worthal a human skeleton (**3**) found lying on the tin ground below a small cairn may be of prehistoric date. On other occasions the finds were discovered at some depth within the streamworks, although their exact context is now unknown. Amongst these are a human tin figure found nine feet below the surface in a streamwork in Lanlivery parish, and at Lanherne, a Late Bronze-Age hoard was found at some depth in the alluvium. However, the great majority of finds from the streamworks, as implied above, cannot be so closely provenanced. For example, two Bronze-Age vessels from Broadwater Moor, Luxulyan; pick axes of stag horn from Carnon and Goss Moor; an 'Iron-Age Tankard' from the Pentewan Valley; wooden pick axes from the Porth streamwork near Tywardreath; spearheads from tinworks at St Hilary, St Erth, Pentewan, Jamaica Inn and Roche; 'celts' (stone axes) from streamworks at Treloy and Carnon; a bronze fibula and jet object from Pentewan and a bronze pin from St Columb.

As early as 1602, Richard Carew first noted that the streamworks were 'very ancient and first wrought by Jews with pickaxes of holm, box and hartshorn; they prove this by...those tools daily found amongst the rubble of such works'. Carew also wrote that

3 A human skeleton found at Perran-ar-Worthal lying on the tin ground below a small cairn may be of prehistoric date. (Henwood, 1873, 206-8)

many bronze axe heads, which he called 'thunder-axes' had been found in some streamworks. The shaft at Pentewan and the antler picks are good evidence for the extraction of alluvial tin during the prehistoric period, but is it possible to determine the character of these operations? The evidence of the shaft at Pentewan suggests that in some instances, underground mining of alluvial deposits was being carried out. Archaeological and geomorphological evidence, however, suggests that this type of operation may have been rare. The tin ground in the nineteenth century varied considerably in its depth below the surface, with for example that at Porth, Tywardreath, being 9.75m (32ft), whilst Pentewan was 13.41m (44ft), Narbo Carnon 16.76m (55ft) and Restronguet 21.33m (70ft). The considerable depth of the tin ground below the surface has led Penhallurick to suggest that the prehistoric tinners may have had frequent recourse to use shaft mining to recover the alluvial tin. There is, however, sound evidence to suggest that the earliest tinners would not have always needed to employ such sophisticated techniques, because the tin ground may have then been at or very near the surface in many valleys. Colin Shell has suggested that the earliest tin extraction would have occurred within the upland stream-tin deposits where the depth of overburden was shallowest and the ore was consequently the most easily available. I.G. Simmons, writing of Dartmoor, noted that tinning doubtlessly encouraged settlers to come to the area with the consequence that the alder, which favoured the damp valley bottoms and was an excellent source of wood for the charcoal needed for smelting, declined disastrously. It would thus appear that the Bronze-Age decline of alder could be an indication of the arrival of the tinners. On Foweymore, pollen analysis of six pollen profiles by A.P. Brown has indicated that a precisely comparable alder decline was visible in level HT 6ai/6aii, which he assigned to *c.* 1100 BC. Unfortunately, this date cannot be accepted unequivocally, since it was based only on correlation with the excavated settlement site at Stannon Down, and its presumed impact on the local flora. However, putting the problems of precise dating aside it is significant that the alder decline coincided with the evidence for a marked increase in pastoral and limited arable agriculture. It would appear that a tinworking and agricultural economy evolved on the moor at precisely the same time, and the two are likely to have been associated in some manner.

Large numbers of prehistoric settlements survive on Dartmoor and Foweymore and it is very tempting to equate the particularly dense settlement pattern with tinworking (**4**). On both moors, prehistoric settlements survive in large numbers near tin streamworks and this has led various writers over the years to suggest that many were built by tinners. We need to be very careful when considering this type of evidence, but it certainly needs to be examined. Some settlements consist of houses with no apparently associated fields or enclosures. Many of these structures may have been sheilings, but it is always possible that some may have been primarily a tinners' settlement or that summer grazing was supplemented with an indeterminate level of tinworking, as perhaps at Witheybrook where the houses are situated near to the alluvial deposits. The settlements associated with enclosures, on the other hand, must definitely have supported a population that practised a pastoral economy. Again, however, the economy of those settlements near to the cassiterite deposits may have been influenced to a varying degree by tin extraction. Finally, those settlements with associated fields were clearly involved in agriculture, possibly with grazing of outfields, whilst tinning was available as an option. Therefore, it is possible to explain the economic character of all settlements without reference to tin and none of them can, on present evidence be interpreted as those solely used by tinners. This situation suggests that the economy may have been a mixed one with all the settlements close to the tin deposits relying on grazing and sometimes arable agriculture to some extent. It is also assumed that there was tin working although there is no positive evidence that this has had any meaningful impact on the settlement pattern.

The leading exponent of a link between the settlement pattern and the tin deposits is David Price whose work on the Dartmoor settlement pattern has convincingly demonstrated that the largest prehistoric settlements are generally closer to the alluvial tin deposits than the smaller examples, and he has suggested that the character of associated field systems may therefore have been directly influenced by tin exploitation. Price also believes that the evidence for a well-developed territorial organisation, as represented by the reave systems, is in part at least a manifestation of a need to support a well-developed tin industry. Andrew Fleming also believes that 'control of tin resources would have been one reason for an elite, whether local or foreign to the area, to take steps to colonise or exercise increased control over these uplands'. Evidence for such an elite is provided by the well-organised reave systems. A connection with mineral exploitation is suggested by the main reave-building episode coinciding with Wessex II in which Camerton-Snowshill daggers, with a high tin content were a predominant feature. It is thus possible that the settlement and field pattern of prehistoric Dartmoor may have been influenced by the tin industry, but clearly further work is needed to ascertain whether the archaeological landscape as a whole may have been influenced by mineral exploitation. Arguments which

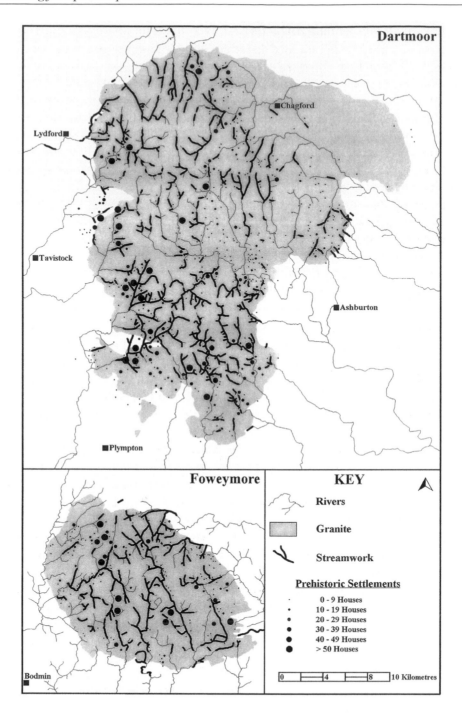

4   *Map of Dartmoor and Foweymore showing the location of tin streamworks and prehistoric settlements. Many of the largest settlements on both moors lie close to streamworks, but this distribution may reflect the obvious need for a good reliable source of water. (Source: Devon and Cornwall Sites and Monuments Registers)*

attempt to demonstrate tinworking activity by examining only the settlement size, character and distribution are fraught with problems. Although a few ideas and interpretations have been suggested above, it must be emphasised that only large-scale excavation of the settlements, combined with a programme of environmental analysis of the nearby tinworks, will ever offer the opportunity to ascertain any correlation between the different types of prehistoric settlement and the contemporary tin industry. The paramount reason for this is that the Bronze-Age settlements of South-West Britain survive in greater numbers than elsewhere in the country and consequently it is not possible to explain the settlement character with purely tin-orientated interpretations, since other differences between this area and the rest of the country are likely to have contributed to the situation. The arguments considered above do at least suggest directions in which future work should be taken.

Further tantalising clues that suggest early control of tin are provided by a small number of sites in Cornwall. At Carn Brea, work by the Cornwall Committee for Rescue Archaeology (CCRA) revealed that some of the southern ramparts may have been deliberately positioned to enclose a particularly rich lode. The evidence to support this suggestion is based on the very obvious re-alignment of a probable Bronze-Age rampart at the point where it would have crossed over the lode, thus excluding it, but instead it took an alternative route (which made no defensive sense) and thus enclosed the lode. This lode may have been particularly rich and thus worth protecting for the sole benefit of the Carn Brea inhabitants. Tinwork protection may have been the primary function of Bolster Bank, which could have afforded some protection to the rich cassiterite deposits of St Agnes Beacon, and another bank in Penwith that ran between Tregenna and Clodgy may have served a similar function.

Excavated evidence for a sound correlation between agricultural communities and the tin industry has come from a number of sites. The recovery of a hoard of tin nodules from a hut at Trevisker and a fragment of slag and a cassiterite pebble from a house at Dean Moor tentatively support the suggested link between Bronze-Age settlements and the industry. At the Iron Age villages of Chysauster, Goldherring and the Roundago of Castallack, all settlements with well-developed field systems, evidence for smelting was found, supporting the hypothesis that mineral exploitation was carried out by communities whose archaeological surface remains alone would point to an exclusively agricultural economy. In the defensive enclosures of Carloggas and Kenidjack Castle, evidence for tin smelting and smithing suggested that one of the functions of south-western hillforts was to act as centres for preparation of the finished products and presumably also for their distribution. This point was emphasised by Threipland, who noted that the siting of the camp at Carloggas must to some extent have been influenced by the availability of stream tin.

Smelting sites were not confined to the settlements and some have been found in or near to the tinworks. Amongst these are Ballowall; Carnanton and Trerank. Smelting at either locality would have had advantages. Those at the settlements would have been more easily protected against criminal attack whilst furnaces at the tinworks would have minimised transport.

The archaeological evidence for smelting consists of small bowl furnaces, consisting of

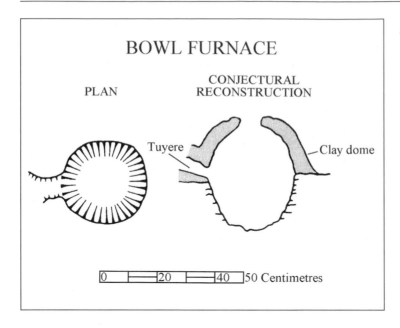

BOWL FURNACE

PLAN   CONJECTURAL
                RECONSTRUCTION

Tuyere   Clay dome

0   20   40   50 Centimetres

5 *A simplified drawing showing a bowl furnace. Furnaces such as these were used during the prehistoric period to smelt tin. The resulting ingots would have been small and plano-convex in shape (After Tylecote, 1980, 210)*

a pit dug into the ground, to receive the molten metal, whilst a flimsy superstructure of stone and/or clay would have formed the furnace itself (**5**). A mixture of the black tin and charcoal would have been placed in the furnace and then burnt to release the molten metal into the bowl below. To increase the temperature within the furnace air was forced in by the use of bellows, or in some instances by the prevailing wind. The resultant ingots would have been small and roughly plano-convex in shape. Large numbers of ingots with this shape have been found and whilst some may belong to later periods because smelting techniques appear to have remained largely unchanged until the thirteenth century AD, some are almost certainly prehistoric. Amongst these are ingots from Penwithick, St Mewan, St Wenn, Praa Sands and Bigbury Bay.

The archaeological evidence for prehistoric tin working thus suggests that alluvial deposits were most often worked and the rich lodes only rarely, whilst the tinners themselves probably practised a mixed economy with mineral-related employment taking up a variable percentage of their time.

It is not until the Iron Age that a minute quantity of documentary evidence is available and whilst this is non-British in origin its very existence indicates that there must have been trade between the literate Mediterranean countries and South-West England. The precise character of the European tin trade has been fully studied by J.D. Muhly, who has identified the many routes through which the tin travelled from Britain to most parts of the continent, but particularly to the Aegean and Mesopotamia. The earliest possible reference to Cornwall as an exporter (and consequently also a producer) of tin comes from Herodotus, who wrote in about 440 BC:

> But concerning the farthest western parts of Europe I cannot speak with exactness, for I do not believe that there is a river called by foreigners Eridanus issuing into the northern sea, whence our amber is said to come, nor have I any

knowledge of Tin-islands, whence our tin is brought . . . This only we know, that our tin and amber come from the most distant parts (Herodotus iii 115. Translated by A.D. Godley, Historiae, 4 vols. (new ed. 1960-61), ii. 141, 143)

This account indicates that Herodotus was not personally aware of the location of the 'Tin-islands', and it is consequently pointless to become involved in the controversy of their precise identification and instead it is noted that Britain is probably the source being referred to. The first specific reference to Cornish tin comes from about 8 BC and is contained within the writings of Diodorus Siculus, who not only confirms the character of the trade but more significantly offers some scanty details of how the mineral was extracted:

The natives of Britain by the headland of Belerium are unusually hospitable, and thanks to their intercourse with foreign traders have grown gentle in their manner. They extract the tin from its bed by a cunning process. The bed is of rock, but contains earthy interstices, along which they cut a gallery. Having smelted the tin and refined it, they hammer it into Knuckle-bone shape and convey it to an adjacent island named Iktis. They wait till the ebb-tide has drained the intervening firth and then transport whole loads of tin on wagons. There the dealers buy it from the natives and thence convey it across to Gaul. For final stage, proceeding by land through Gaul, in about thirty days they bring their load on horses to the mouth of the Rhone. (Diodorus Siculus, V.22: source J.D. Muhly, 1969, 472-3)

The extraction technique described above would appear to be mining. The earthy character of the material being worked would indicate that this does not refer to rock mining, but rather to shafts or pits dug into the alluvial deposits in a manner suggested by the archaeological evidence from Pentewan. The contemporary classical documentation confirms that tin was being produced for the European and Near Eastern markets in the late prehistoric period and this combined with the archaeological evidence for earlier links, supports the hypothesis that British tin was traded from the Bronze Age onwards.

The Roman period

The arrival of the Romans in Britain in the first century AD does not seem to have been accompanied by the increased activity in tin extraction which one would have expected. All other mineral industries flourished and expanded as a consequence of the Roman conquest. The gold mines at Dolaucothi in Wales, lead and silver mines in Derbyshire, North Wales, Shropshire and the Mendips and copper in Shropshire, North Wales and Anglesey were all opened or expanded. However, in South-West England the tin would appear to have been left largely unworked in the first two centuries AD. The most plausible explanation relates to the known supremacy of the north-west Spanish mines. There are only two known exceptions to this situation. First, is the evidence from the tinwork at Boscarne where a variety of first-century objects have been found. Significantly, near to this site is the only Roman fort in Cornwall and it is tempting to see this as being established to facilitate extraction of the nearby tin, under military direction. Parallels for this situation are to be found at the gold mines at Dolaucothi and the lead mines in the Mendips, and suggest military control of the mines, with the work being carried out by slaves. The tinwork and associated fort at Boscarne were abandoned in the later first century, presumably when it was recognised that no minerals other than cassiterite were to be found. Second, excavations at Castle Gotha revealed evidence of second-century clay-lined pits, hearths and a mould for casting pennanular bronze bracelets. The cassiterite used at this manufacturing site was probably streamed from the nearby rich deposits of Pentewan. In approximately the middle of the third century, with the exhaustion of the Spanish mines, British tin once again became an important natural resource, needed by the pewter industry at Camerton (Somerset) and Lansdown (outside Bath) for the production of jugs and dishes. Further uses were made, and the widespread wearing of bronze brooches in particular, and tinned bronze jewellery in general must have been responsible for the increased production.

The archaeological evidence for renewed tin exploitation is diverse and the different elements are considered below. The most significant evidence is three tin ingots, each of which Beagrie has demonstrated were of late Roman date. The one from Carnanton bears two or possibly three impressions, one of which is an inscription reading 'IENN', which unfortunately is not yet understood. The second piece of evidence results from the work of Brown et al. whose analysis of sediments from the River Erme suggests that the valley bottom deposits were brought into production during the third century AD. The third is from an excavated hut on Par Beach, St Martin's, Isles of Scilly and at Trethurgy Round another ingot was found associated with Late Romano-British pottery. Less convincingly, an adit at Baldue was believed by Hunt to be Roman because of the 'perfection of the arch formed and the good masonry of squared stones of which it was constructed'. Much more secure are: the tin pan and cover from Treloy, Colan, (**6**) which were found on undisturbed tin ground; a tin bowl dug up at Parson's Park in the immediate vicinity of a tinwork; in the parishes of Paul, Ludgvan and Towednack many Roman coins have been found; in Marazion Marsh (a likely location for a streamwork) an earthern pot was discovered containing a thousand copper coins of the emperors who lived between the years AD 260 and 350; and more specifically Norden, noted that the Romans 'took their turn to work for tin, as is supposed, by certain of their money found in some old works renewed'. Wacher believes that the Roman Villa at Magor, Illogan, may have been directly associated with the tin industry — perhaps as a residence of the wealthiest Cornish Master-tinner — because of its uniqueness in the landscape and proximity to rich tin deposits. There is no direct evidence to support this suggestion, but one would certainly have expected the creation of an administrative elite within the industry and it is just possible that this site reflects this situation. Further indirect evidence for the third-century reopening of the Cornish tinworks comes from a small number of inscriptions and milestones that could suggest an attempt to build roads to facilitate the export of tin from the area. Despite this later Roman exploitation of tin, with the possible exception of the Villa at Magor, the industry does not seem to have had any particular impact on the settlement pattern, with for example there being no settlements created to serve and

6 *The tin pan from the streamwork at Treloy is described as being "discovered in Tinground which had been previously wrought". (Henwood, 1873, 221)*

administer the activity and certainly no mining towns comparable to that at Charterhouse in the Mendips. This suggests that the industry was not centralised, and the work must have been carried out at the local level, with very little military control.

The Dark Ages

The evidence for tin extraction in this period is again somewhat scanty, although there is sufficient circumstantial and some more positive indications to suggest that it continued. The most secure evidence for working comes from Trewhiddle, where the complete treasure of an ecclesiastical community including a number of coins inscribed with dates from AD 757 to 901 was found buried in the debris of an old streamwork. This find strongly suggests that streamworking had been carried out in the area in the ninth century, although the possibility of the hoard being buried in a much earlier and long disused work should not be ignored. In addition, a second find of a probable fifth-century penannular brooch from a streamwork on the Goss Moor may point to tinworking in this area. At Colliford, analysis of a pollen profile collected within a streamwork suggested that it had been abandoned around AD 600-700. The fourth and final piece of site specific evidence for tinworking comes from Chun Castle where a furnace and a twelve-pound lump of tin ore were found in a sixth-century context. This has led Leslie Alcock to conclude that: 'It seems very likely that control over the mining and working of tin provided a main source of wealth'. The circumstantial evidence comes from the fact that tin was being widely used throughout this period either for fine jewellery and on a larger scale for church bells and consequently because these materials were used implies a trade in metals, and suggests also the existence of specialised communities engaged in exploiting the ores. There is a relatively substantial body of documentary evidence demonstrating the continued exploitation of Cornish tin in particular during the Dark Ages, and since this has already been collated and reviewed by Hatcher, all that need be noted here is that it confirms the archaeological evidence that the South-West continued to produce tin, both for the British and European markets. Given the evidence for continued cassiterite exploitation and trade, it is perhaps surprising that there is no reference to tin in the Domesday Book

(1086). Lewis felt that the possible reasons for this were that the Stannaries were considered a Royal property and were, therefore, not subject to taxation, or that Danish and Norman attacks may have forced their temporary closure. H.P.R. Finberg, on the other hand, believed that the Stannary revenues were so insignificant that they were included with the miscellaneous revenues. Without the evidence to support the acceptance of any of these suggestions, it must simply be noted that tin does not appear in the Domesday Book, but that production is certainly known for the periods both before and after.

3 The medieval and early modern period (AD 1066 — c. AD 1700)

The existence of abundant contemporary documentation for this period makes it possible to examine the many characteristics of the industry, which cannot be studied by fieldwork techniques alone. Details of the whole process can be studied, and we now know how much tin was being produced, by whom and where. The legal framework in which the industry operated is also understood and taken together, a comprehensive view is available. Thanks to the work of Lewis, Hatcher and Pennington the complex, economic and legal aspects of the industry are already well understood and are therefore not repeated here. Instead, the emphasis is on examining and explaining the processes involved in producing tin. In the next five chapters the different stages and processes involved in extracting the tin ore from the ground and converting it to the finished ingot will be examined using both archaeological evidence and contemporary documents

This section examines the processes involved in setting up a tinwork. First, the different ways in which the valuable cassiterite was found are considered. Second, the legal processes in staking a claim are examined and finally some of the methods employed to raise the necessary capital are discussed.

Prospecting

Prospecting was a crucial activity within any tinwork. Before extraction could start, it was essential to find out as much as possible about the location, character, quantity, quality and extent of any tin deposits or lodes. The process did not stop once exploitation had started because crucial to the long term survival of any tinwork was the need to continually find new material worth extracting. The success or otherwise of a tinwork could, therefore, depend very heavily on the prospecting skills of the tinners. A rich lode not found could be as catastrophic to the tinners' fortunes as a poor lode wrongly identified. Given the importance of prospecting, it is therefore not surprising that both the documentary sources and surviving archaeology have a great deal to tell us about the different techniques employed and the myths that developed surrounding that especially rich find. For many tinners, the chance of finding that special lode or deposit was all-consuming. Sadly, the chances of finding it was probably less than winning today's lottery and worst of all, with each year that passed the odds grew longer. Stannary Law revolved around the concept that even the humblest of tinners could become wealthy by finding a rich deposit and working it profitably for their own benefit. This probably encouraged adventurous individuals to prospect in the hope of bettering their position, but success was probably relatively rare because of the economic realities that favoured capital intensive ventures.

The earliest known British account of prospecting is provided by Norden, who in 1584 wrote that:

> both the Streame and Loade-workes, are founde by litle stones, which lye both in, and nere, the Brookes, and upon the mountaynes, wher the mettall lyeth: And theis Stones they call the Shoade, being parcell of the veyne of owre, which being dismembred from the bodye of the Loade, are means to directe to the place of profite, as the smoake directeth where the fire lurketh' (Norden, 1584, 12)

This description is not particularly helpful since it seems to imply that the tinners relied exclusively on finding the telltale 'shoade' on the surface. Life for the tinner was somewhat more complicated. The process of prospecting generally involved a great deal of hard work, which frequently resulted in no gain. The archaeological evidence suggests that the most common prospecting method was the digging of small pits, as many thousands survive throughout the stannaries (**7**). This is confirmed by the documentation with many writers providing us with their accounts of the processes involved. The earliest reference is provided by Richard Carew, who wrote that:

> To find the load works, their first labour is also employed in seeking this shoad, which either lieth open on the grass, or but shallowly covered. Having found any such, they conjecture by the sight of the ground which way the flood came that brought it thither, and so give a guess at the place whence it was broken off. There they sink a shaft, or pit, of five or six foot in length, two or three foot breadth, and seven or eight foot in depth, to prove whether they may so meet with load. By this shaft they also discern which was the quick ground (as they call it) which moved the flood, and which the firm, wherein no such shoad doth lie. If they miss the load in one place, they sink a like shaft in another beyond that, commonly farther up towards the hill, and so a third and a fourth, until they light at last upon it.' (Carew, 1602, 90)

Carew's optimistic account of prospecting is clearly the product of a county gentleman who had never experienced the backbreaking task of excavating even the smallest of holes. Despite this, his undoubted knowledge of the industry cannot be denied and he does provide us with a useful insight into the methodology of prospecting using pits. Other writers mention prospecting methods and amongst them are the anonymous writer of 1670 who refers to them as 'Essay Hatches' and the technique as 'trayning', Pryce who called the practice 'costeening' and Agricola who is uncharacteristically brief on this matter, simply called the pits themselves 'trenches' and gave no name to the technique.

The most complete account is that provided by an anonymous writer who at some considerable length describes the different stages of the process. From his account, four major pieces of information may be gleaned:

7 *Contemporary illustration showing dowsing and the digging of prospecting pits. (Agricola, 1556, 40)*

1. Establish the presence of shoad, which having been weathered from a nearby lode, was the evidence that trayning could be profitably employed.
2. Sink an essay hatch into the ground at the lowest available point, examining all the material removed, for shoad. A careful note was made of the depth at which the shoad occurred because the lower in the profile it was found, the closer was the lode. This pit was dug to the depth of what the writer called the 'shelf', which was presumably the bedrock.
3. Further pits were then sunk progressively up the hillside, and the process repeated until the shoad was found lying directly on the bedrock. This indicated that the lode was near at hand and if in the next pit no shoad was found, the lode would be located between these final two pits. More pits were then dug to reveal the lode and its character.
4. Complications were encountered where there was more than one lode on a hillside, but these could all be found, because the shoad from each would have occurred at a different depth, within each pit, and the skilful tinner would thus have been able to establish how many lodes were responsible for the phenomena observed in the hatches (Anon., 1670, 2096-2101).

One might properly expect examples of series of pits excavated in the course of trayning to survive in the field. In both counties, many examples have been found and amongst these are a fine series of pits leading toward the lode-back tinwork at Hobb's Hill (*see* **40**), the pits around the eluvial streamwork west of Black Rock (**8**) and at Black Tor where pits were used to locate and then examine the lodes (**9**). The pits themselves survive individually as small rectangular or oval hollows with an associated crescent-shaped bank,

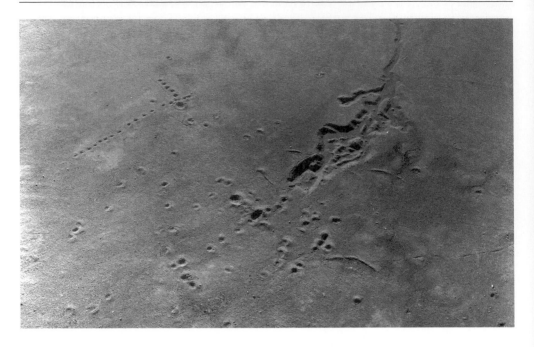

8   *This small eluvial streamwork west of Black Rock on Foweymore is associated with several reservoirs and series of prospecting pits. The systematic manner in which prospecting was often carried out is clearly illustrated by the lines of pits. (Photograph by Steve Hartgroves, Cornwall Archaeological Unit: copyright reserved)*

normally lying downslope of the pit. Carew and the anonymous writer both provide us with descriptions of these pits. The anonymous writer says that they are six-foot long by four-foot broad and as deep as the bedrock, whilst Carew describes them as 5-6ft in length by 2-3ft wide, and 7-8ft deep. Clearly the precise size of the pits varied, but in general terms they appear to have been relatively small. This information is particularly useful in identifying prospecting pits in the field, where sometimes it is difficult to differentiate between this type of pit and the larger pits associated with lode-back tinworks. A recent study of a tinwork at Black Tor revealed that after careful measurement it was possible to identify areas of prospecting and extraction (**9**). In this instance it was also possible to demonstrate that after the lodes had been identified they were subjected to further examination by the use of pits dug along their length. This discovery highlights the dangers in assuming that all pits dug onto the back of a lode were used for extraction.

The second major prospecting method that has left an obvious mark involved the use of water to remove much of the overburden, in a manner similar to the hushes employed in the lead industry. The anonymous writer noted that gullies, similar to those produced by flooding could be made in likely areas by cutting a leat:

> about 2 foot over, and as deep as the Shelf, in which we turn the water to run
> 2 or 3 dayes: by which time the water, by washing away the filth from the

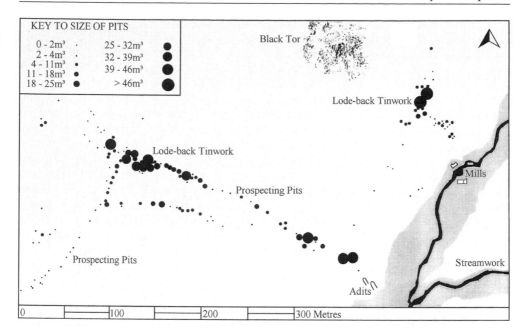

KEY TO SIZE OF PITS

0 - 2m³	25 - 32m³
2 - 4m³	32 - 39m³
4 - 11m³	39 - 46m³
11 - 18m³	> 46m³
18 - 25m³	

Black Tor

Lode-back Tinwork

Lode-back Tinwork

Prospecting Pits

Mills

Prospecting Pits

Streamwork

Adits

0   100   200   300 Metres

9 *Simplified plan showing the distribution of prospecting and lode-back pits south of Black Tor. The areas of lode-back working are highlighted by the largest solid circles, whilst a clearly defined series of prospecting pits can be traced leading towards the largest lode-back pit tinwork. The tin extracted from this tinwork was probably crushed at the nearby stamping mills*

stones, and the looser parts of the earth, will easily discover, what Shoad is there' (Anon., 1670, 2098)

A series of at least twelve trenches produced by this kind of activity survive at Hart Tor, where the water was carried in a 650m long leat from a nearby river to a reservoir, before being released to cascade down the hillslope (**10** & **Colour Plate 1**). On a smaller scale, a single gully leading down a steep slope from a leat in the Yealm Valley at NGR SX 61806360 and another overlooking the North Walla Brook at SX 68068338 are probably more typical. More unusual are the sites where wholly hand-dug trenches were excavated instead. Examples of this type survive at Goonzion (**33**), and Kit Hill (**34**) where water was in short supply.

Another prospecting technique was the digging of adits (or drifts) through likely areas to try and find the lodes at depth. On Hingston Down, in 1617, Nicholas Harris was granted 'liberty to drive and work the adit now being driven home on the loade'. In 1659, Thomas Bushell proposed the cutting of a two-mile long adit through Kit Hill to intercept any lodes that may have been present. Although this work was not carried out, the very fact that such an ambitious scheme was contemplated demonstrates that this technique was far beyond its infancy. Finally, in 1690 the adventurers at Ball West 'carried our Additt to finde and sink Loads as we had cutt in our drift'.

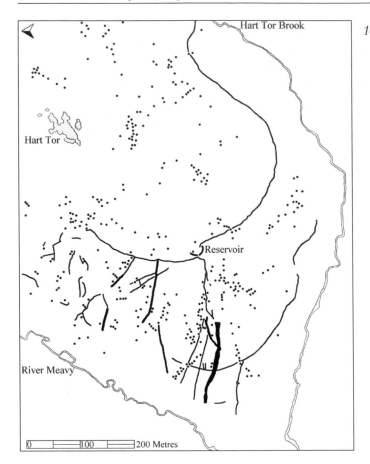

10 *Plan of prospecting features around Hart Tor. The small circles represent prospecting pits whilst the lines are leats and prospecting gullies. The leats and gullies are cut in a number of places by later prospecting pits, although in at least two locations the gullies would appear to divert around pre-existing pits. This would suggest that some prospecting using pits was carried out first, before much larger scale investigations using water were contemplated. Finally, a further programme of exploration was carried out using pits*

The final intrusive prospecting technique would have been the cutting of leats to serve the tinworks and mills. Though primarily excavated to carry water, they would have also performed the secondary function of providing the tinners with an opportunity to examine the underlying geology. There are hundreds of miles of leats in the tin producing areas and the number of lodes stumbled upon in this manner may have been considerable.

As well as the prospecting methods that have left their mark on the landscape, there are a whole range of techniques which we only know about from the documentary sources. Amongst the most imaginative of these is the use of dreams to find the ore. Presumably once the dreamer reached the site of their nocturnal thoughts the more mundane task of examining the area would have been carried out by costeening or trenching. Carew's faith in this paranormal technique is abundantly clear and to emphasise its validity he recorded that:

> one Taprel, lately living, and dwelling in the parish of the hundred of West called St Neot, by a like dream of his daughter (see the luck of women) made the like essay, met with the effect, farmed the work of the unwitting lord of the soil, and grew thereby to good state of wealth. (Carew, 1602, 90)

Pryce, also briefly mentioned dreams but was much more cynical of their value, writing that 'we have a Huel-dream in every Mining parish, which raises and disappoints by turns the sanguine hopes of the credulous adventurers'. In a similar vein, a legend concerning Goonzion Downs in St Neot parish suggests that tinners sometimes relied on saints to find them their tin. On one occasion St Neot is said to have promised to show the local tinners the location of a particularly rich lode by placing a feather on it. Superstition was probably an inevitable characteristic of the early tinner, given the nature of the work in which chance rich discoveries could be encountered at any time without there being any apparent logical explanation.

Another technique used to find the lodes was examining the ground following its disturbance. Thus the anonymous writer noted, that after flooding, any gullies which formed, were examined 'to see, if happily we can discover any metalline stones in the sides or bottoms thereof'. This writer also noted that sometimes the shoad was found in mole hills or during ploughing. In addition, to these, Agricola, noted that gales, accompanied by the widespread uprooting of trees, earthquakes, avalanches, forest fires and poaching by hoofed animals could all be employed in the search for minerals. Using only areas of existing disturbance would necessarily have limited the tinners considerably, and in those parts where these phenomena were scarce, artificial techniques were employed, to achieve the same result.

According to many writers, lodes could be discovered without the need to disturb the surface. The anonymous writer, Borlase and Agricola all noted that vegetation growing immediately above a lode was of a very different character to the surrounding flora. Usually, the vegetation above a lode is stunted and this is probably because it has been partially poisoned. An example of this phenomenon is visible above the Tinners' Lane cross lode, St Neot, where aerial photographs highlight the position of the lode as a narrow much lighter band, in contrast to the darker textures on either side. Agricola noted that hoar frosts were less likely to occur over lodes, and that consequently they could be easily distinguished under the appropriate conditions. Henwood recorded that water flowing through lodes was often 'slightly heated' and Collins, J.H. noted that in metalliferous areas, springs frequently indicate the position of a mineral outcrop. If these factors were understood by the early tinners they may have contributed occasionally to the discovery of lodes.

The most often non-intrusive mentioned prospecting technique is that of 'divining' or 'dowsing'. This method involved holding a V-shaped branch of hazel firmly between both hands and walking over the ground until the rod deflected, when the dowser passed over a lode. Agricola (7), Pryce and Leifchild considered this technique at length and believed seriously in its value. Henwood, recorded that the practice 'was formerly much in vogue amongst the Cornish miners' and Leifchild wrote that it 'has been in frequent use, and often mentioned since the eleventh century'. It is, therefore, likely that many of the lodes found in early times were discovered by this technique, and their character subsequently proven by the digging of prospecting pits along their length. In summary, many tin lodes were found by first searching for, and then locating the shoad which either lay at the surface because of some form of natural disturbance or was revealed by one or more of the many techniques employed to examine the shoad at depth. Other lodes were discovered by examining associated natural phenomena, dowsing and even good luck. The numbers found by each of the techniques will never be established, but all probably played some part in the history of the industry.

So far, we have examined the archaeological and documentary evidence for prospecting lodes. Clearly, tin deposits in the valley bottoms were also prospected before streamworking began. Here the principle of digging a pit to examine the character and quality of the deposit was the same, but in this situation the process of trayning was not needed because the tin deposit would have been confined to the valley bottom. The result is that most of the prospecting was carried out within the valley bottom using pits known as hatches to examine the tin deposit. On occasions, prospecting pits are visible on the edge of the valley bottom, but in these circumstances they would have been excavated either as part of trayning to search for a lode on the adjacent hillside or to find the edge of the deposit. Large numbers of hatches in streamworks are known and amongst these are the substantial ones at Colliford (**11**) and a solitary example at Lydford Woods (**22**).

Historical documents and the surviving archaeology together confirm that accurate prospecting was an essential part of any tinwork. Without the information revealed by this work, it would not, in most cases, have been possible to start the process of extraction. Adventurers would only commit their capital or time to a tinwork with promising prospects. A tinwork might, therefore, wither or flourish on the quality of its prospecting tinners.

Bounding

Having found a potentially worthwhile cassiterite deposit or lode, the practice was then to stake a claim by cutting or pitching bounds. This involved defining the area in question with a pile of turfs at each corner and, from the late fifteenth century onwards, this was followed by registering the claim with the appropriate Stannary Court. The regulations regarding bounding were continually being altered and since these subtle changes have been documented by Pennington, they are not repeated here. It is sufficient to note that the earliest Stannary Charters of 1201 and 1305 allowed bounds to be pitched anywhere throughout the county. However, by the sixteenth century the tinner had to inform the landowner of enclosed land that he had bounded a particular area, and before work could commence permission had to be granted. This change in regulations is best illustrated by two examples of grievances. In the first, dated 1361, John de Treeures petitioned the Black Prince complaining about the tinners who were destroying his crops, whilst by contrast in 1650 six tinners were charged in the Blackmore Stannary Court with trespass at Dowgas because it was claimed that they had worked for tin within enclosed land without the landlord's permission. The evidence in this latter case was solely concerned with proving that the land had been enclosed prior to being worked for tin, and the subsequent failure to demonstrate this resulted in the acquittal of the defendants. Proving ownership of bounds was important to avoid disputed claims, and this was presumably why Prince Arthur's Council of 1496 insisted that:

11 *Part of the alluvial streamwork at Colliford, highlighting two hatches and their shared drainage level. The leat leading to the southern hatch was carried on an embankment and approached the hatch at such a height that it could have powered machinery*

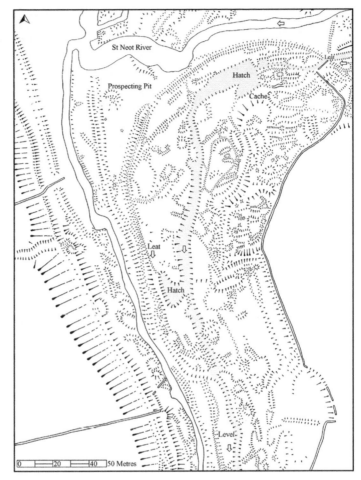

If any tinner shall hereafter pitch any tinwork he shall at the next law court enter the whole bounds of the same tinwork and the name of the tinworks with the names of his fellows... and the steward or his clerk shall take for every such entering but a penny for every name'. (Pennington, 1973, 81 citing B.M. Add. MS. 6317)

The introduction of bound registration in 1496 explains the apparent proliferation of documented tinworks in the sixteenth century, since before this date individual examples are rarely mentioned in the Stannary Court Rolls. Later in 1588 these measures were further tightened, with renewals having to record both the old and new names of the work. The effect of these actions can be seen in the format of the surviving documents which give details of former names (if applicable) and relatively detailed descriptions of location, making it possible to identify these bounds accurately. One particular example relates to a streamwork on Penkestle Down (**12**):

One paire of Tynnworke Bounds formerly called Arrows Flight and now called

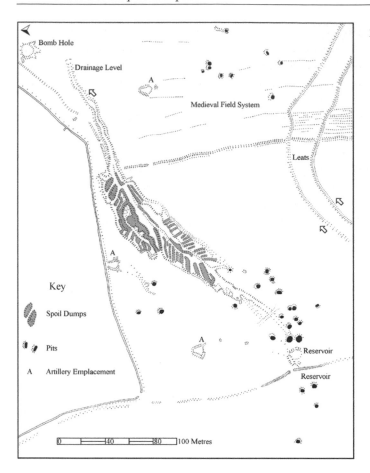

12 The eluvial streamwork on the northern side of Penkestle Moor is on the site of a tinwork known as Arrows Flight. This name may have been derived from the chevron shape of the earthworks in the southern part of the tinwork. The artillery emplacements and bomb hole date to World War II and highlight the dangers of using fieldwork evidence alone to explain archaeological features. The bomb was dropped over a year before the emplacements were constructed

by the name Goodfortune Bounds cutt and pitcht by Thomas Courtis the Seaventeenth day of March in the yeere of our Lord 1690 to the use of John Cole gentl. John James gentl. and Thomas Courtis the younger the North East Corner Lyeth nere an old Leate the South East Corner nere the River Lyver otherwise Colliver River the South West Corner Lyeth nere an old dead headge or bound And the north west corner Lyeth nere the same old hedg in or uppon certain Lands in or called West Colliver in the parish of St Nyott. (R.I.C. Ma/3/6)

The probable reason for the stricter control of bounding was to prevent disputed claims, which could be so wasteful, but these still occurred, with for example, Thomas Cossan, William Gregor and Bennet Spargo being involved in a dispute in the late seventeenth century, with George Tremaine concerning 'Tyn Bounds' near Kregbrears, in Kenwyn. That it was possible for different tinners to claim an area previously bounded by others indicates that bounds lapsed if they were not renewed. To prevent such loss the owners were obliged, at the very least, to maintain the bound markers, whilst in the sixteenth century in Penwith and Kerrier Stannary the minimum level of expenditure to prevent forfeiture was set at the equivalent of three months cost for one man. Tin bounds figure significantly in the documentation and it is thus possible to comment on their ownership and value. Hatcher has noted that despite the Stannary conditions apparently favouring the small-scale operator, the industry was controlled from the twelfth century by a relatively small group of wealthy entrepreneurs. This situation was certainly confirmed for the later period with large quantities of bounds being owned by the wealthier members of society. Thus for example in 1668 Stephen Hickes and Thomas Cossan jointly owned shares in 35 pairs of bounds; in 1691, Hugh Tonkin owned shares in over 170 bounds (mainly in St Agnes) and in 1633, Mr. Sawle owned shares in 44 Blackmore bounds. The value of bounds varied considerably, depending on whether they were either in full production, perceived to be potentially rich, exhausted or simply barren. It is impossible to place a value on many bounds as they were commonly sold in groups, but a number of examples, all from the seventeenth century indicate the variety of prices paid. In 1682 Philip Hawkins purchased John Harris's share of the Fatt Work bounds on Dowgas Common for 5s; whilst in 1684 on St Hilary Downe the White Worke Bounds which were working were valued at £35, whilst the nearby Tinkers Bound was worth only £1. However the true difference in value of bounds may be judged by comparing the £10 paid by James Praed for shares in seven tinbounds, in 1673 with the £11 paid five years later for an eighth part in the single pair of Wheale an Gours, St Agnes (total value was probably about £88). These relatively small-scale transactions represent the majority of bound speculations, but on occasions substantial amounts were paid for what must have been extremely rich examples. Thus in 1688 John Cothy of St Austell sold his share in sixteen tinbounds to John Hayman of St Mawgan for £845 and in 1672, £550 was paid to Wm. Randell of St Ewe for shares in tinbounds, Polgooth adit, tin at the mills, the stamping mills themselves and 'utensils'.

Tin bound owners were thus often people of considerable wealth and are described in the documents as gentlemen, merchants and tinners. These people did not limit their interests to single areas and thus for example Thomas Courtis owned shares in tinbounds at North Bounds and Good Fortune at Colliford and at Good Luck and Breedwell in Perranzabuloe. John Foote owned shares in bounds in the Stannaries of Blackmore and Tywarnhaile. In the late seventeenth century Thomas Hawkins owned bounds in Blackmore and Penwith and Kerrier and the Carlyon's of Tregrehan owned bounds in all the Cornish Stannaries. This diversification was not limited to owning tinbounds over a wide area, but also in having non-stannary related interests. Thus John Cole of Cartuther esq. who owned shares in tinbounds 'wthin the Severall pishes of St Nyott, Poundstock, Blisland, St Kewe, Cardinham, St Breoke, St Agnes, Redruth, Lanivett, Stythians, Gwinnap, Illugan' also owned large areas of property and farming equipment. Possession of bounds did not relieve the tinners of the obligation of paying Toll Tin (or Gadered Tyn) and on some occasions Land Dole to the landholder as compensation for the damage to his property. The legal situation regarding the collection and right to Toll Tin and Land Dole is complex. However, in Cornwall, the landlord had a claim to a percentage of the tinwork's produce (Toll Tin) and after 1686, a share in the profits (Land Dole). Whilst in Devon, the landowner was less well served by the Stannary Law and only possessed a claim to a share of the net profits. This meant that if the operating costs were greater than the proceeds from the sale of the black tin, the unfortunate owner received no compensation. Richard Carew recorded that the Land Dole was normally one fifteenth, but according to Pennington, on Foweymore it was 'reciprocal of the number of the adventurers shares plus one'. This must have encouraged some adventurers to group together in as large a number as possible to reduce this burden. Some particularly adventurous landlords commuted their right to Toll Tin in return for a share in the tinworks. Thus for example, in 1582 at Trewolvas, Hugh Trevanyon awarded a Sett to Edmund Trevithicke in return for a third share.

The amount of Toll Tin paid by the bounders to the landowner varied considerably, with one fifteenth (e.g. at Lanyon in 1632) at the lower end and one quarter (e.g. at Russells Worke in 1673) at the upper. This was probably a consequence of negotiating the level of payment. For example Sir William Godolphin on being approached for a grant to worke in Chycarne and Tresowes' opened negotiations by offering:

> that they should when they brought home the Audditt to Ball West have the preferrence of a sett therein att a reasonable dish & that hee (Sir William Godolphin) would not put in any other before them in regard the said Audditt may draw the same day. (C.R.O. DD.J. 1337)

The different levels of Toll Tin levied probably reflect the anticipated wealth of a bound, with the richer ones paying a higher percentage than the poorer ones. This situation is illustrated within the manor of Treverbyn, Blackmore, where the levels of Toll Tin payable on different parts of the property were clearly stated at the Blackmore Stannary Court meeting on 25th April 1616:

> in Kerrowe More from ye bounde stone betwene Trevanyons land called Chenowyth & the tenement called Rosveare, Easteward to a lane called old Wykes lane, leading from Rescorla, the tenth pte of the tynn there founds is dewe for toll, And ye West Syde of ye stone — All the waste of that manner payeth ye thirteenth to toll except ye pte of the worke called Burlazack al's Bolazack wch is belowe ye craze, on ye Southe as fawr as the bunds thereof extend wch pte onlye payeth ye tenth. One payer of bunds called ye greate Beame payeth ye fyfteenth pte the workes called the lyttle Pentrouth + Great Pentrouthe on the southe syde of Rouses hedge als osgveare payeth the tenth, The rest of that worck payeth the thirteenth, one worck called Carbusse Worck payeth the tenth pte and all except great beame payeth ether the tenth or the thiteenth pts for toll. (R.I.C. HEND. 14.306)

This example is not unique, and on a similar smaller scale, in 1686, Jonathon Rashleigh of Menabilly tried to encourage further extraction at the 'old work' at Higher Work Park, Ennys, by levying a toll of one sixth compared to the higher level of one fifth charged for the nearby new work. Another approach adopted by landowners to encourage production (and hence profit for himself) was to reduce the Toll Tin charge in return for a share in the bounds. In 1687, such an adjustment was made by John Janner of Court esq. who:

is by custom of mannor to have the tenth part of all Tin broken in the manor and is ready to promote the advantage of the Tynner + Tin affairs. He agrees that if all Tin Bound owners in Goonbarne Goverseth Drinick Hard-come-by Stenaguin will grant a fifth part to him he will only demand one thirteenth doll instead of one tenth (R.I.C. HEND. 18.59).

Acceptance of this proposal would have meant that Janner was more directly concerned with the net profit rather than solely output, and it would thus have been in his best interests that these bounds were efficiently managed to increase his own interests and those of his fellow adventurers. In some instances the Toll Tin was levied on the 'head' (the best) and 'tayles' (the poorer tin) separately and at West Park, St Austell the charge was a 'fifth part of all the head tin found there and also one fifth of tayles of the said tin as shall be left, more worth than twelve pence a foot' (C.R.O. DD.CN. 2526).

In the seventeenth century, the matter of Toll Tin was regularised into the practice of landowners granting Setts to adventurers, whereby the general conditions as well as the level of payment were agreed. The legal position regarding the different forms of Sett is very complex, but in summary there are two main types; licence and grant. The rights of the adventurers given a licence were limited to exploiting the minerals, whilst a grant gave them a legal interest in the land. The latter Sett type was the most secure and enabled the adventurers to, for example, legally prevent other parties entering their tinwork, whilst in the former the grantee relied upon the landowner for such protection. The Sett licenced to the adventurers at Ball West, Germoe is a good example of the disadvantages of this type, for when 'Mr Pennecke sent severall Tinners that denied our workeinge + made several pitches before + behinde us tho tis within our sett' their only course of action was to write to Sir Williame Godolphin (the landowner) asking him 'to order in Mr Pennecke to give us not farther disturbance'.

Pennington has noted that the acquisition of bounds by non-working proprieters began in the early sixteenth century, though his evidence is slight, consisting as it does of the assumption that Richard Code a wealthy landowner was unlikely to have cut his bounds on Foweymore personally. However, by 1539 the practise had been accepted in at least Penwith and Kerrier and by the seventeenth century it was commonplace with for example: Thomas Holmes, John Jenkin and Sampson Rose cutting bounds for John Foote and others. Lewis Ellowe, in 1675, cut bounds at Wheal-an-Venton for the use of William Boscawen Esq. and Hugh Bawden, gent. In 1676 bounds in Kenwyn belonging to Bro. Grenfeild and Stephen Hickes were renewed by 'Edward Ostler, Roger Ostler, Humphrey Ostler, Wm. Ostler, Tho. Saunders, + his sonn William, Symon Tregra + John Peeters al's Harris'. On Midsummer Eve 1665 St Margitts Bounds 'was renewed' for Mr. Sawle 'by Ma Thomas + Edwards and the Tynners that worke in St Margitt' and in 1674 Lewis Tremayne paid John Bone 15s. for 'cutting Bounds in Gover and Trewoone'. The widespread use of employees to cut bounds reflects both the highly capitalist nature of the industry and confirms that by the seventeenth century ownership of bounds had escalated to such an extent that it was no longer possible for the individual owner to renew them personally. Tinners either having bounded an area and obtained the necessary permission from the Stannary Court and landowner or having agreed to work a Sett already bounded by the granter next had to ensure that they had sufficient capital for the venture.

Collecting the necessary capital

The need for, and method of raising the necessary finances to initiate the exploitation of bounds is most clearly outlined by Carew who wrote:

> When the new found work enticeth with probability of profit, the discoverer doth commonly associate himself with some more partners, because the charge amounteth mostly very high for any one man's purse, except lined beyond ordinary, to reach unto; and if the work do fail, many Shoulders will more easily support the burden. These partners consist either of such tinners as work to their own behoof, or of such adventurers as put in hired labourers. (Carew, R., 1602, 91)

The existence of many names on some of the bounding certificates clearly indicate that the practise of tinners associating with partners was often carried out at an early stage, but often these shares (or doles) were further subdivided as work proceeded, with the consequence that individuals often only owned a very small interest in any particular bound or tinwork. For example, in fourteen bounds claimed by Mr Bonithon in 1673 the fractions involved were eight at an eighth, one at a fifth and five at a quarter. In 1570 at Clennacombe Streme and beme worke the names and share of ownership of eleven people are known. Whilst in a Sett taken up at Porthkellis in 1691 details of the nine shareholders involved together with their responsibilities to the venture are recorded. This practise ensured that the risks were spread and probably prevented the wealthier elements losing their money on their combined tin interests, and may have allowed some of the poorer tinners to risk investing in individual tinworks, where they could never have hoped to own their own complete venture. The subdivision of ownership may thus, to some extent have encouraged the wider-scale ownership that the Customs of the Stannaries were geared to encourage. The amount of capital needed to exploit a particular cassiterite deposit would have varied considerably, with the large mines being the most expensive and the small streamworks the cheapest. Some of the smaller streamworks may have been exploited entirely by family groups, perhaps with the additional help of a few full-time or part-time labourers. It is, unfortunately, impossible to comment on the extent of this level of working, since the documentation so necessary for the efficient management of the larger concerns, generally never existed in the same quantities for the smaller.

The major method of attracting finance to a tinwork was to offer shares for sale or to borrow money against the security of doles. The first of these measures involved the encouragement of merchants, gentry, affluent tinners and wealthier yeomen to invest in the tinwork for a proportional share of the profits, and with this capital the necessary equipment and labour could be purchased. The second alternative, borrowing, was probably only the last resort, since the interest rates were often extortionate and the Stannary Court Rolls reveal frequent examples of moneylenders gaining shares in tinworks as a result of default on payments. A successful tinwork would have been a much sought-after property and this would have increased the value of the doles and consequently also the amount of capital invested in it. This in return, given the appropriate geological conditions, could further increase its value and associated profits. In 1558 Thomas Roydon of Truro, merchant, sold three-and-a-half doles in eight 'yn a Certain Tynworke callyd the Pell' to John Cosowarthe of London, Esquyer, for £300, and if it is assumed that the remaining doles were of a similar value, this tinwork must have been worth around £728. Another valuable tinwork was The Tye in St Stephen-in-Brannel, for in 1684, a twentieth dole was sold to Thos. Carlyon of St Austell, gent. for £48, giving the whole work a value of approximately £960. The justification for the assumption that each dole within a tinwork had a similar value comes from 'the myne of Polgouth' where in 1598 and 1601 'one full quarter of a dole yn 14 doles' fetched 20 nobles (£7.1.8) and £6.13.4 respectively. The reason for the decrease in the value of the doles at this tinwork over three years may have been caused by a decrease in the value of tin as a consequence of the enormous increase in Cornish output from 790,950lbs. to 1,067,112lbs. Utilising these figures the value of the whole tinwork can be calculated at between £375 and £398. The transfer of shares was not always amicably executed and in 1676 Peter Allen, St Austell, tinner claimed that Sam. Blake, St Austell, tinner, tricked him into selling his shares in St Margarets and Mulberay during a 'drunken discourse'. On other occasions, shares were acquired by providing a service rather than money. Consequently Thos. Roydon and others received two doles and the fourth part of half a dole in seventeen doles for bringing 'an audyte or tye' into the Pell, St Agnes. Many shares were bequeathed either to the church or the individual's family. Thus in 1493 William Martyn of Lostwithiel left 'one and a half Doole Stanner called de Quelhiocke... to the priory of St Andrew of Tywardreth' and Dole Stannar called Le Halzebet 'to the Church sancti de Luxilion'. Tho. Tregarthen's will of 1508 left his daughter Philipp his shares in tinworks at Mulfra, St Austell, Polgooth, Goss Moor, Crugebras and elsewhere provided that John Chamond and Jane Poyle 'shall govern the said work until she and her husband be better of abilitie to govern and gyde them', and Richard Trevonans in his will of 1552 left 'ys wife the 9 part of my work yn Parke an Fenten to Jane Chilton my god daughter the like. . . To Renold son of Alson my daughter the 8th part in Whele an Fenten. To Willm. my brother and Robert my son the dole of every work I have except Whele an Prat'.

These wills are particularly significant because they indicate that women owned tinwork shares and in the case of Jane Poyle she was probably actively involved in their management. Ownership of the tinworks was generally limited to the same group who owned the tinbounds, and although wealthy gentry, merchants, tinners and yeomen possessed the lion's share, poorer elements of society were also represented. The Probate inventory of John Hore, Roche, 1589, indicates that he was not a particularly wealthy farmer, but he also owned 'righte in certaine tynworks' with a value of £6.13.4 and Walter Agoe, Luxulyan, 1603, of similar status possessed a right in a tinwork worth £1.

4 Tinworks and tinbounds

Having found the tin deposit or lode, bounded the work and raised the necessary capital, the next stage involved extracting the cassiterite. This was achieved using streamworks to exploit the tin deposits and mines to quarry the lodes. In chapters five and six the details of the different techniques, conditions and character of the surviving archaeology are considered. Within this chapter a number of themes common to both major types of tinwork will be examined. Amongst these are employment conditions, the water supply, a perusal of the evidence relating to the ancillary buildings found at many sites, an insight into reasons why tinworks were abandoned and sometimes reopened and finally an examination of tinwork distribution.

Employing and paying the work force

From at least the thirteenth century many tinworks were capital orientated, with the shareholders either paying money into the venture or providing an employee or employees, instead of working the claim. Consequently in 1357 we find that Abraham the Tinner employed over three hundred men; John Thomas, in the early fifteenth century owed his employees twenty marks; in 1674 Lewis Tremayne 'pd the Tynners' who worke in Trewoone Moor' £17.9.8; in the later seventeenth century Poldice, Gwennap, employed 800 and 1000 men and boys, John Ayer and his partners employed 'servants and workmen' at Stertmore and in 1697 Celia Fiennes wrote that within twenty mines around St Austell more than 1,000 men were employed. These examples indicate that throughout the later medieval and early modern periods, individuals were employed in the tinworks, and it is possible utilising sixteenth-century documentation to ascertain that there were up to three categories of employee.

Thomas Beare writing in 1586 noted that there were three types of stannary worker, of which two were employees:

> I take the tynner to bee him that giveth wages by the yeare to another to worke his right in a tynworke for him as a doale or half doale more or less, or else worketh his Right himself as many doe. The Worker is hee that taketh uppon him to serve the Tynners Right for wages by the yeare or else for less tyme. The Spalliard is hee that cometh Journeys now and then to ye Tynworke for his hire and for the day saveth the worker from spall as the Tynners terme it, and so by that occasion is called a Spalliard. (B.M. Harleian 6380)

R.H. Worth's work on a variety of Cornish documents added a further category and more importantly details of average wages. According to Worth the three types of employee working in the sixteenth-century tinworks were firstly, spalliards who were men working by the day, or journeymen, receiving 2s per week, and casual labourers taken on at the wash and other times. Secondly, labourers who either had yearly contracts with a fixed wage, but no interest in the results, receiving not more than £4 10s per year. Also, Dole-workers were given a part share in the mine for the year or half-year and took part of their master's profits, but also worked a part share for his benefit and for that were paid a reduced wage, usually 20s a year. Thirdly, an intermediate class who took half tin, half wage. The wage was usually half of £4 10s or thereabouts. This class became actual partners in the works for the year. Worth divided the employers themselves into two categories. The first were the tinners who took over at a rent a share in the work. The rent was usually stated as a percentage of the tin raised, for example the fifth dish or such like. Second, there were the master tinners — who owned shares in works, but either employed labourers or set their rights, taking no personal part in the labour.

Thomas Beare was not silent on the question of wages and he noted that 'the common wage is but iii li of five marke a doles working for the year to the uttermost'. Carew, writing a few years later, observed that: 'The hirelings stand at a certain wages, either by the day, which may be about eightpence, or for the year, being between four and six pound, as their deserving can drive the bargain, at both which they must find themselves'.

The division of workers into groups is not apparent within much of the documentation. The general term 'tinner' is employed to describe every type of employee, many of the owners and anyone involved with the industry. The sixteenth-century labouring tinner was notoriously poor, and it was probably the employees who were most at risk, since they were numerous, had no personal investments and consequently the owners could keep their wages low and still rely on obtaining sufficient labour. This situation was made much worse by widespread usuary and the development of the 'truck system' whereby part of the loans were paid in goods. The extreme poverty of the employee class of tinner is best illustrated by Beare who wrote:

> This poor man happely hath a wife and iiij or v small children to care for which all depend upon his getting, whereas all his wages is not able to buy himself bread, then to pass over the poor mans house rent, clothing for his poor wife and children besides divers other charges dayly growing upon them. O god how can this poor man prosper . . . (B.M. Harleian 6380)

The picture was, however, not completely bleak and Beare went on to note that the wealthier tinners were often generous to their fellow workers:

> As the riche tynners will lack none being lest of them in numbre, then is their charitie so great that if one two or 3 or els more poor men syt among them having neither bread drink or other repast there is not one among all the rest but will distribute at least the largest sort with the poor worke fellowes that have nothing, so that in the end this poor man having nothing to relieve him

at work shall in the end be better furnished of bread butter cheese beef pork
and bacon, then all the richest sort'. (B.M. Harleian 6380)

In addition, the tinwork owners sometimes provided financial help for their employees,
with for example John Eva in 1633 promising 'to give my tinners some small content for
their aide until it come to more plentie'. Despite these ameliorating factors the tinners' lot
remained harsh, leading in some years to riots and other disturbances. By contrast, the
moneylenders, who were often also the tin merchants, tinwork owners and landowners
made considerable profits at the expense of the working tinners whose low wages and
consequent susceptibility to usury, meant that they too were exploited along with the
cassiterite. Given these prevailing conditions, Hatcher observed that 'Tinning was held by
many to be an occupation into which one was driven by necessity rather than choice'.

Capital Outlay

The documentary and archaeological evidence for the different types of tinwork is
considered fully in Chapters 5 and 6. It is worth noting here that the size, type and success
of a particular venture depended initially on the richness and accessibility of the cassiterite,
and then on the ability of the tinners to raise the necessary capital to exploit the mineral
efficiently. The largest and more inaccessible tin deposits and lodes required greater amounts
of capital to maximise profit, whilst the smaller quantities and readily accessible material
could be economically extracted on a shoestring budget. Thus, for example, the relatively
deep but rich lodes at Ball West, Germoe, could only be reached after digging adits to lower
the water table, and this would have only been possible if the adventurers had sufficient
capital to pay the employees for this necessary preliminary work. Preparing to bring a mine
into production could be costly, with there being no mineral output, yet substantial
overheads. Consequently success often depended on there being sufficient capital available
to meet these initial costs. The late seventeenth-century adventurers at Ball West emphasised
this point in a letter to Sir William Godolphin: 'we have beene att great Cost with out a
penny proffitt since we began this'. Failure to find a plentiful supply of ore before the initial
capital was exhausted may have forced the premature closure of many mines. If, however, a
worthwhile lode was found, the extraction of this could then have paid for its mining and
dressing as well as the initial development costs. After this stage was reached the tinwork
would probably continue to function providing that the operating costs were offset by the
value of the tin recovered. Eventually however, as a result of a number of possible
circumstances, costs would have edged above the crucial level at which the operation could
economically survive, and consequently closure would inevitably follow.

Reworking

Tinworks were often brought into production many times. There are three main reasons
for this. First, advances in technology may have made tin ore available that previously

could not be reached. Second, an increase in the price of tin may have made it economical to extract ore that had previously been uneconomical. Third, a tinwork may have been previously abandoned through no fault in the quality and quantity of output, but rather as a consequence of personal circumstances such as the death of an adventurer, failure of capital or a dispute between the involved parties. Improved technology affected the viability of disused tinworks in two ways. First, developments in extraction techniques enabled the tinners to extract tin that could not be reached previously. Second, improved processing ability facilitated the extraction of ore, which had, until this time, been impossible to collect. In the former, the introduction of efficient drainage, for example, would have enabled the miners to reach depths of ore previously abandoned as a consequence of either insufficient technological ability, or the lack of finances to put the known technology into practise. Hamilton Jenkin has comprehensively documented the eighteenth- to twentieth-century history of the many mines that were being continually opened and abandoned. This work has demonstrated that a combination of technological inexpertness, problems with finances, and disputes between partners were the major reasons for closure. These may also have been important in the early tin industry, since documentation certainly survives to indicate that old works were reopened.

In 1686, Jonathan Rashleigh of Menabilly granted a licence to a group of adventurers 'to search for tin in a field called the Higher Work Park on the tenement of Ennys in St Enoder with liberty to drive or cleanse any adit to an old work'. A few years later, in 1697, Rashleigh granted Richard Rundle of St Enoder a licence to examine two old disused tinworks at Harvenna. In 1612, Christopher Pollard cut tin bounds called 'Kild Nest' where there had formerly been a tinwork called 'Litle Carburlye'. The same year, John Myle took possession of 'Blacketor Combe' which was 'voyde for want of Renewing of bounds', but on this occasion the new owner was satisfied with the old name and it was not changed. A tinwork on Trevedda Hill, Warleggan seems to have had a particularly chequered history being known as 'David's Church, Brode open worke and Good Fortune'. Multi-named tinworks are probably the result of a work being abandoned and reopened under a new name. This site may thus have had at least three separate phases of extraction by 1613 (the date of the document). These examples form a small sample of the many documented instances of tinworks being periodically abandoned and reopened. That reworking was widespread is supported by Carew who noted:

> There are (those), that leaving these trades of new searching, do take in hand such old stream and load works as by the former adventurers have been given over, and oftentimes they find good store of tin, both in the rubble cast up before, as also in veins which the first workmen followed not. (Carew, 1602, 90-91)

Carew's comments deal with both varieties of technological improvement, for he suggested that tin could be economically extracted from tinners waste, as well as lodes which had not been previously exploited. Working of waste heaps is only worthwhile if the available processing technology can be utilised to extract enough tin to pay for all the costs incurred and provide a surplus. A combination of two factors are responsible for ensuring that later tinners could make a profit from earlier waste.

Improved processing techniques may have meant that tin that could not be efficiently extracted before could now be readily collected, and/or an increased tin value meant that more time and resources could be expended on the waste and still enable a profit to be made. Reworking of waste could thus be made cost effective and Merret and Agricola, both noted this phenomenon:

> The Causalty (ie. waste) they throw in heaps upon Banks, which in six or seven years they fetch over again, and make worth their labour. But they observe, that in less time it will not afford Metal worth the pains; and at the present none at all. (Merret, 1678, 952)

> There are some people who wash over the dumps from exhausted and abandoned mines, and those dumps which are derived from the drains of tunnels. (Agricola, 1556, 30)

The potential of waste is further emphasised by Charles Trewbody's purchase of 'leavings of tin' with the Boscundle stamping mill in 1706. The practice is not one confined to the early period, and Pryce noted that at Perran Arwothall in the eighteenth century 'some hundreds of poor men, women, and children' were involved in collecting tin from the wastes of the upstream mines. More specifically, in 1754 John Parnell of St Austell, tinner, was granted a lease by John Sawle 'to carry to the mills all the tinstuff lying on the heaps or burrows near Carclase Ball, and to stamp, work and dress it'.

The reworking of tinworks must have had an effect on the archaeological record. At many alluvial streamworks, the complexity of earthworks is a consequence of partial reworking and the nineteenth-century phase of alluvial streamworking revealed the presence of possible prehistoric tinworks in many Cornish valleys. Personal considerations for the abandonment and reopening of tinworks are much more difficult to evaluate, since they were probably effected by the individuals economic motivation and psychology of the adventurers.

In conclusion, there is abundant evidence to demonstrate that the extensively documented eighteenth- to twentieth-century practice of mines being constantly taken out of and brought back into production, and their dumps being reworked, was also a phenomena of the early tin industry

Cessation

The exploitation of a particular cassiterite deposit was only possible in a capital-orientated industry, if the value of the output was greater than the monetary input. The only exception to this situation was during an optimistic exploratory or development phase, when the adventurers would have been willing to operate at a considerable loss in the hope of later compensatory gains. However, in a mature mine in which profits were slowly dwindling and there was no perceived likelihood of finding further reserves of economically worthwhile mineral the most probable outcome was the closure of the operation before the adventurers lost money. There were many factors that could transform a profitable tinwork into one which made a loss. The first was that as a tinwork developed, and the more accessible ore was exhausted, it became increasingly more expensive to reach the rest. Thus for example, ore above the water table could be worked without the need for pumps, but with the deepening of the operation it would have needed drainage devices and these would have inevitably raised the operating costs. At a certain depth (which would have varied from mine to mine) the additional costs of pumping and handling of the ore to grass (the surface) would have increased to such an extent that ore below this level, although possibly as rich, could not be extracted economically, and the mine would have closed with reserves untapped. Evidence to support this point comes from a Lansdowne manuscript of 1575 and this states that:

> the tynn works do decaye, not onely for the contynuall workinge in the same whereby the worke is more deeper and therefore more chargeable to come by, but most chieflie for that the Tynners fyndinge small comoditie thereby forsake the said worke and fale to Tyllage. (B.M. Lansdowne MS. , xviii, fol. 52)

Ore not exploited by a particular phase of mining could possibly be extracted later when the operating costs relative to the mineral value were more conducive to exploitation. This situation was, probably more than any other, responsible for the documented phenomenon of tinworks being reopened and bounds being maintained after the cessation of full-scale mining. It also explains why much of evidence for early mining and its associated processing has been destroyed, since the reopening of these tinworks in modern times often removed traces of previous activity.

A second potential cause is the variability in tin value, with a higher price being responsible for making previously unviable deposits a worthwhile proposition, and a lower price of course making deposits that had been economic no longer so. Work by Lewis has shown that the price paid for tin varied considerably through time and from one market to the next. Thus from one year to the next the economic viability of a mine may have hung in the balance. When the price was high, large profits would have been made and if in the following year the price plummeted, the losses could have been so high as to force its closure. The price paid to the adventurers would have also varied according to the quality of the tin and perhaps more importantly their ability to negotiate a favourable recompense. Thus if two mines produced tin at the same cost, one may have thrived because its adventurers were in a position to offer their tin at the highest available price, whereas the other may have closed because its owners were in debt to merchants who were consequently able to force them to sell their tin at a lower uneconomic price. Fluctuations in tin price combined with either the ability or inability to receive the maximum available return may have been responsible for the closure or success of many mines.

A third factor responsible for cessation of mining was the exhaustion of the ore. The Lansdowne manuscript of 1575 (cited above) indicates that this was a major cause of desertion. With the removal of all the mineral, the inevitable consequence was an end to extraction. Many of the lode-back tinworks and openworks on Dartmoor and some in Cornwall show no trace of modern mining and may, therefore, have been completely exhausted at an early date. However, this phenomenon was probably responsible for very few closures, because most of the tinworks have been reworked in modern times indicating that some deposits had been left, and in these cases it was probably the exhaustion of the economically viable ore that was responsible.

A fourth factor that may have had some limited effect was desertion of the tinwork by the employees, presumably either to return to agriculture or to look for better payment at other operations. Hatcher has noted that this was a particular problem in the later fourteenth century, when labour was in short supply.

Another possible reason for abandonment may have been accidents such as flooding or collapse, which proved difficult to overcome. At Relistian, near Gwinear, a rock fall in 1681 killed 24 men and closed the mine for a time. Accidents would have been particularly damaging to the small-scale operations, where such unexpected problems could not be overcome financially and were thus more likely to prove economically disastrous.

A serious dispute between partners in a tinwork could also have been catastrophic, with money and energy that should have been invested and expended at the tinwork being utilised instead at the stannary courts.

Finally, failure to have a Sett or lease to work extended may have caused a premature end to some tinworks, since without this permission extraction could not have continued. A combination of these various factors was probably responsible for the closure of tinworks and although the precise conditions would have varied from one venture to another the cessation of work would have inevitably involved the dismissal of the workforce and the selling off of the equipment. All that remained were the physical traces of mining and any associated buildings, which were later, if they survived, to become an invaluable source of archaeological information.

Water Supply

Water was crucial for many activities associated with the tin industry. Streamworks, mines, and mills needed access to it in varying degrees and it was even used during some prospecting operations. In most situations considerable effort had to be expended to get the water where it was needed and this has left a considerable impact on the landscape within the stanniferous regions. There were three main sources of water available to the tinners.

Direct from a river

Only alluvial tinworkings were able to use water without the need to carry it in artificial channels, and even in this instance, only a limited amount of work could have been carried out. When tin was first found in a valley, rich deposits would have been available to the tinners who would have simply panned for it within the existing river. With the exhaustion of the deposits within the riverbed, the next stage was probably to throw the material from the river bank and surrounding area into the water. This technique would have been the earliest in most tin bearing valleys and inevitably has as a result left no trace in the archaeological record.

Leats direct from a river

This water source involved building a weir or dam across a river and diverting the water into a leat that carried it to its destination. Leats are commonplace and can be traced running along the hillsides, often for some considerable distance. The tin mills all needed water to function and in every known case leats were employed. A large and constant supply of water was needed by processing sites since it was the prime source of power and essential to dressing.

Among the longest known leats are the South Hill and Bradford leats, both of which clearly demonstrate the extraordinary hyrdological skills of the tinners. The 8km (5 miles) long South Hill leat is considered to have been cut during the latter part of the fifteenth century to supply tinworks at Teigncombe and a mill at South Hill. The Bradford Leat which carried water for about 19.3km (12 miles) from Watern Combe to a tinwork at Bradford Pool was cut at a cost of £1,500 in the early part of the sixteenth century and fell into disuse at the end of the seventeenth century after a series of court cases which the tinners ultimately lost.

Leats from springs and run-off channels

This source is much more unreliable than tapping a river or stream, but in some circumstances only springs and temporary storm run-off channels were available. A particularly complex example lies on Letter Moor **(13),** where a series of leats collected water from the western side of the moor and Searles Down. This water was fed into two large V-shaped reservoirs situated on the watershed between the St Neot and Dewey valleys and then carried by more leats to the Southern Penkestle eluvial streamwork. This scheme was ambitious and demonstrated the lengths to which the tinners were willing to go to extract the available cassiterite in areas where water was in short supply, and the extent of the remaining streamwork earthworks is a clear indication of the success of their efforts. Another site with an impressive array of leats is Stanlake where a combination of spring and river fed channels were constructed to serve a relatively small eluvial streamwork. More typical of the leats serving eluvial streamworks are those at Harrowbridge and St Lukes. In these instances a single leat was cut to carry water from the springs and storm flow channels to reservoirs on the higher ground above the streamwork.

Reservoirs in which water could be stored are often found in association with leats. These were necessary to provide a more reliable locally available source of water. The reservoirs vary considerably in size and shape, but all include a bank behind which water was held. Most reservoirs are associated with eluvial streamworks.

Ancillary Buildings

Shelters

Many tinworks were in remote and exposed locations and consequently shelters were needed in which to rest from the arduous task **(14).** The existence of these shelters is confirmed by the contemporary documentation. John Norden wrote in 1610 of 'many

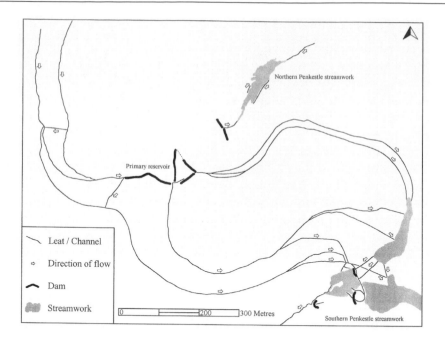

Northern Penkestle streamwork

Primary reservoir

Leat / Channel

Direction of flow

Dam

Streamwork

0   200   300 Metres

Southern Penkestle streamwork

13 The hydrological skills of the tinners are clearly demonstrated at Letter Moor. A complex system of interconnected leats and reservoirs illustrate their capabilities. Much of the water for the Southern Penkestle streamwork was brought across the watershed from a neighbouring valley

litle howses buylte for the Stannerie men to shrowd them in neere the worckes', and Thomas Beare noted in 1586 that:

> at dinner tyme... they syt downe together beside theire tynwork in a litle lodge made up with turves covered with straw and made about with hansome benches to sit upon (B.M. Harl. 6380)

Specific references to shelters are unfortunately lacking, but it is possible that the 'old tynn howse' referred to in the 1613 bounding of Perranuthno and the two 'Tin houses' at St Margaretts, in 1668, may have been used as shelters. This shortage of contemporary accounts is not surprising since stannary documents are unlikely to have been concerned with such a mundane and, more significantly, unproductive aspect of a tinwork.

By contrast the archaeological record is much more illuminating, because associated with many tinworks are small rectangular or sometimes triangular buildings (**Colour Plate 2**) which have been identified as the shelters recorded by Norden and Beare (**Colour Plate 3**). These structures vary considerably in size, but on average those on Dartmoor have internal dimensions of 4.84m long by 2.7m wide (16 x 81/2ft), whilst those on neighbouring Foweymore are similar, with an average 5.14m length and 3.1m width (17 x 10ft). They are defined by a crude dry-stone or turf wall often with a surviving doorway, sometimes a fireplace and more rarely a cupboard. Some shelters are built

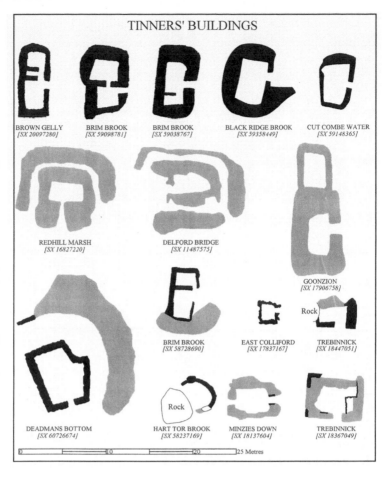

14 *Simplified plan showing the character of a number of tinners' buildings. Where the walls are shown solid this indicates that the building appears to be entirely composed of stone. The shaded structures on the other hand survive only as earthworks and may be of turf construction. Excavation of the building at Redhill Marsh confirmed that it was entirely composed of turf*

against large earthfast rocks, others, especially in Cornwall, have a ditch around their upper side, and many, lie within earlier prehistoric round houses. A small number of reused round houses were excavated by the Dartmoor Exploration Committee, but the results were not particularly informative. The excavation of a turf built shelter at Redhill Marsh in the 1980s was only partially more successful with a few sherds of pottery suggesting a sixteenth-century date. The majority of the buildings could have sheltered between 10 and 20 tinners giving a valuable insight into the maximum number of tinners which could have been employed at any one time in the individual tinworks.

Many of the tinners' buildings lie within tinworks and only survive where reworking has not destroyed them **(Colour Plate 4).** A large number of the earlier shelters were probably destroyed in this way.

Storage Buildings

There are no general accounts that describe buildings being used for storage, but specific references indicate that, at some tinworks, a building was put aside for this purpose. Thus at Clowance in 1684, 16 gallons of black tin was being kept in the 'Tynn House'. At Twelve Heads, in 1613 there was a 'newe menerall house' and at Ennis, St Enoder, in the late seventeenth century the adventurers were permitted to construct a 'material house or houses'. The physical character of these particular structures is not known, but it is likely that they were built to serve the anticipated need, and are thus likely to have varied considerably from one venture to another. Many small structures found within and close to tinworks are interpreted as storage buildings **(Colour Plate 5)**. They are often referred to as caches and they survive as very small chambers often composed of drystone walling attached to a rock outcrop or earlier feature such as a wall, whilst others are dug into earlier spoil dumps. They are often circular or oval in shape and are only big enough to hold tools or perhaps black tin (a concentrate of the cassiterite). The anonymous writer confirmed that black tin was kept in a safe place, for he noted that it was 'put up into Hogsheads covered, and lockt till the next blowing'. The only cache known to have been excavated was at East Colliford where a small drystone structure cut into an earlier waste dump was examined revealing no dateable artefacts and no traces of occupation **(14, Colour Plate 6)**. At least 205 shelters and 40 caches are recorded on Dartmoor, whilst the number in Cornwall is thought to be similar. Given the small size of these structures and the relative infancy of archaeological fieldwork it is considered likely that many more await discovery.

Accommodation

The mining settlements so prevalent in the nineteenth-century industrial scene are completely absent from the earlier period. In the moorland areas, where many of the tinworks lie, the predominant settlement pattern was one of dispersed or partially nucleated farmsteads and none of the tinworks are associated with structures that can be unequivocally identified as the homes of tinners. On Dartmoor, in particular, an examination of the distribution of streamworks and medieval settlement **(15)** clearly illustrates that there does not appear to be any obvious relationship. Clearly some settlements do lie near to streamworks, but many more do not and there is certainly no obvious clustering of settlements to take account of rich accessible tin deposits. The primary reason for this is that the focus of working within each area would have moved constantly and consequently there would have been little point in establishing a settlement relying on a resource that would have been worked out after only a short time.

The question as to where the tinners lived is linked to the hotly debated discussion between Blanchard and Hatcher on the status of medieval tinners. Both parties agree that many 'miners' possessed land. However Hatcher believes that 'it is fallacious to proceed from the proven fact that many miners possessed some land to the conclusion that mining communities were self-sufficient as far as their food supplies were concerned', whilst this was the very approach favoured by Blanchard. The intense nature of the published discussion between these two historians, combined with the fact that both were able to provide strong evidence to support their respective arguments suggests that both situations were to be found within the British mining scene. The prevalence of one situation over the other may have varied with time, by season or even from place to place, but it is not within the scope of this book to examine afresh the intricacies of tinner status.

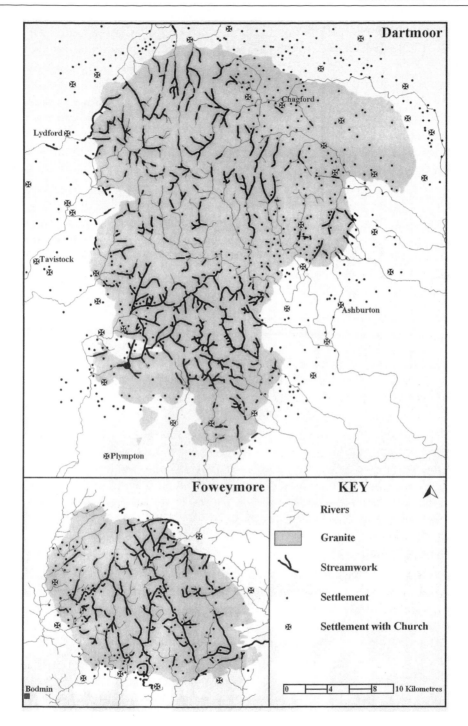

15 Distribution map of Dartmoor and Foweymore showing the position of tin streamworks and medieval settlements. Only those settlements on the granite and adjacent areas are shown. Most of these streamworks would have been exploited during the medieval period. (Source: Devon and Cornwall Sites and Monuments Registers)

51

The significance of this complex situation is that each different type of tinner would have lived in accommodation that reflected their position within the industry. The gentry, with stannary interests, would have occupied their ancestral homes, and because they were not involved exclusively with the industry and did not play a physical role in the extraction process, their accommodation was often a great distance from the works themselves. Thus, for example: Jonathan Rashleigh of Menabilly was involved in the tinworks at Ennis, St Enoder, some 18km (11 miles) distant from his residence; Sir James Smyth of Chelsea was, obviously, even further from his work at Pouldice, and William Scawen of Molenick possessed interests in a considerable number of tinworks on Foweymore. These people must have employed managers to oversee their varied and diverse stannary interests, and consequently the distance from their works was not as disadvantageous as such a situation would have been to the working tinners. Examples of tinworks being managed on behalf of third parties is abundant with for example: Thomas Hawkins' bounds being, 'renewed and looked after' by Arthur Gavid; whilst Mr Sawles' bounds were renewed by a variety of people, including Richard Handrake, Edward Harris, Mr. Thomas and Mr. Edwards. In 1695, a group of bounds were 'cutt and pitched ... by John Paynter and George John for the use of Charles Bodvile Earle of Radnor, Sir John St Aubyn Barronett, Alexander Pendarves Esq. and Edward Chapmen, gent'. and in 1508, John Trevergan was responsible for collecting the toll tin for St Perys Egecomb. Profits from the gentry's tin ventures could, however, have effected the physical character of their residences. Income from the tinworks may have been invested in new, additional or improved buildings and thus contributed to their general splendour.

The merchants, another group with considerable stannary interests, also lived some distance from the tin production centres, being largely confined to the ports or coinage towns. Thus, for example: Bryan Rogers of Falmouth and Thomas Worth of Penryn owned shares in the Hellanoon tinwork, Lelant in 1683; in 1627; Philip Harris of Clements possessed 'fower doles in 24 in one paire of bounds called Resethdew in Kenwen'; in the second half of the seventeenth century Thomas Cossen and Henry Gregor of Truro owned shares in many tin bounds, especially in Kenwyn. The income from these ventures probably contributed to the building and/or upkeep of spacious townhouses.

The final and numerically largest category is that of the working tinners. Blanchard and more particularly Hatcher have shown that those involved in the extraction and processing of tin were not of a uniform social status. It is thus to be expected that the character of their accommodation would have varied accordingly.

Residences of full-time labouring tinners

The character of the dwellings inhabited by full-time labouring tinners probably varied considerably depending on the individual circumstances of particular tinners. However, given the documented poverty of many labouring tinners it is likely that they generally occupied the poorer quality housing. Unfortunately, fieldwork has failed to reveal examples of houses occupied by this class of tinner, and it is therefore not possible to be precise about their character. On the question of identifying the location of their dwellings, it is possible to suggest a number of alternatives. The full-time tinners relied to a greater extent than the other classes on the purchasing of foodstuffs, and this may have meant that the majority lived within or near to nucleated settlements, where such produce was more readily available. If this was the case, it would certainly explain the dearth of relevant archaeological information, since these structures would have been more susceptible to destruction by later redevelopment and expansion, particularly during the nineteenth-century mining boom. A further argument to support this contention is that the labouring tinners' position of employment at any particular tinwork was hazardous because the venture could fail at any time, or a change of ownership could have left him unwanted. Consequently, it probably made more sense to live in a nucleated settlement situated a relatively short distance from a number of tinworks, rather than immediately next to a single operation. Being positioned centrally would have had the additional advantage of allowing him to choose between the different options, and hence the opportunity to maximise either wages or job security.

Thus far the evidence for occupation of nucleated settlements is largely circumstantial. However documentation certainly indicates that some tinners were in this position, with for example: in 1662, William Blarnie of St Mewan, tinner, leased a dwelling house and garden in Dadeporth, Veryan; in 1633, Edoner Prater of St Blazey, tinner, purchased a cottage and garden in Buscoveyan for £4 and in 1658 Walter Waren of Tywardreath, tinner, leased a tenement in St Blazey churchtown. Thus, an unknown percentage of full-time tinners probably lived within small, nucleated settlements. However it is likely that the situation was much more complicated, and it is known from local folklore that there was some temporary migration to the tinworks, and these people may have lodged with local farmers or other tinners. Another possibility is that some of the poorer tinners being unable to afford their own home may have grouped together and leased a single dwelling. A single possible reference to confirm this practise comes from St Agnes where in 1684 Nicholas Langdon, tinner, purchased an eighth of a house in Mingoose, although it must be emphasised that it cannot be proven that he actually lived here.

In conclusion, the evidence for the character and location of full time tinners' dwellings is very ephemeral and until further work is carried out the picture must unfortunately remain unclear and largely speculative.

Residences of Part-time Tinners

This class of tinner only worked for a variable percentage of their time at the tinworks and supplemented their income either by labouring for the more wealthy farmers or tending their own small land holdings. Sadly, there is no conclusive documentation to say that tinners were employed in agriculture, probably because the work was carried out on a casual basis and records are thus unlikely to have survived. The documentation regarding the second practice is plentiful, although the numbers of tinners involved could not be established.

The first point which must be emphasised is that the job titles are in some instances interchangeable. Thus in 1697, John Tinner of St Blazey was described as a labourer, but by the following year his status was recorded as tinner and between 1674 and 1675, Peter Allen of St Austell was described both as tinner and yeoman. This flexibility of employment titles confirms the existence of a group of people who engaged in other activities besides tinning. The reason for the confusion regarding status is probably because these people stated the type of work in which they were employed at the time of the document's preparation. This practice was, however, not adopted by everyone and may explain the situation, in 1686, of Nicholas Woon of St Enoder, yeoman, being granted a share in a Sett at Ennys, in the same parish. Nicholas clearly considered himself as predominately involved in farming, but with an additional interest in the tin industry. The others (above), on the other hand, may have divided their efforts more equally between the tinworks and agriculture, and consequently described themselves as either tinners or yeomen/labourers. Examination of documents reveals this practise to have been common on medieval Dartmoor where Fox has shown that a few Duchy farmers appear in a list of tinners. Having established that part-time tinners were also involved in agriculture, it is worthwhile examining the character of their holdings.

At the lower level, their property may have consisted of a house, a barn and a few acres. Examples of this type of holding are numerous and include: William Hext who occupied 15 acres (6ha) on Dartmoor; William Wopson of St Blazey, tinner, who in 1681 leased a house with garden and right to common pasture on nearby Tregrehan moor for pigs and geese; in 1686, John Rowse of St Blazey, tinner, leased a dwelling house, barn, back yard and garden with commons of pasture in Tregrehan 'townplace' and 'wastes'; in 1691, Thomas Hoskyn of St Agnes, tinner, leased a house and 3 acres of adjacent land near Mithian and in 1689, Michael Pickett of St Agnes, tinner, leased a house and waste land, with the liberty to cut turf for fuel. These dwellings were situated within the stannary districts, although not necessarily next to any particular tinwork. The amount of time and effort expended on the different activities would have been dependent on the size and quality of the land holding and the availability of worthwhile industrial employment.

Many people described in the documentation as tinners owned or leased more than a small holding and they may be seen as belonging to the same class of worker as represented by Nicholas Woon who was referred to as a yeoman. They would have either split their efforts equally between agriculture and mining or possibly only dabbled in the industry, describing themselves as tinners, solely for the benefits which such status bestowed. It is hard to conceive that Walter Trubody, of Lanlivery, tinner, who leased 60 acres (24ha) of Souther Pencows, would have had much time to devote to tinning. Equally, John Ennor of St Agnes, tinner, who leased 7 acres (3ha) of land and a grist mill at Mithian in the second half of the seventeenth century, probably relied on these activities to a greater extent than on his professed occupation. Details of the type of property held by this class of tinner are significant since they illustrate a certain prosperity, traditionally denied. Hence, in 1674, Andrew Hodge, senior, of St Agnes, tinner, leased houses, chambers, cellars, barns, stables, ground rooms, gardens and land with parlour and room above the hall, half- the barn and mowhay and some wood at Trenoweth. The part played by this group in the industry cannot be accurately assessed, but in terms of settlement pattern, their dwellings would not have been significantly effected by the industry. Since the primary resource was the land, they would have sited to maximise its potential rather than that of the nearby cassiterite deposits. Archaeological excavation of any of these settlements would probably reveal evidence for an agriculturally based economy with barns and byres providing the conclusive evidence. The excavator would then be left to suggest, on the basis of nearby contemporary tinworks, that mineral exploitation may have played some undetectable part in the site's economy, though the scale or even existence of this activity can never be proven. Thus, for example, at the excavated deserted farmstead of Bunning's Park (**Colour Plate 7**), although no direct evidence was found to link it with the nearby Colliford streamworks, it is possible that this upland settlement sited on relatively poor unsuitable soil, survived only because of the additional income offered by part-time streaming. If this suggestion is accepted some of the settlements in the stannaries may be considered to be inextricably linked with the industry, despite a paucity of archaeological evidence. However, since the occupants were not solely involved with tinning, these settlements cannot be seen as the forerunners to the nineteenth century mining ghettos, but instead depended on both agriculture and tin for their creation and survival. To many people involved with the tin industry, agriculture played an important, often primary, role, and this is reflected in the stannary settlement pattern of dispersed and nucleated farmsteads and occasional villages which were geared predominantly to exploiting the agricultural resources, with tinworking playing a secondary, although probably important part.

The primary reason for the observed settlement pattern is that land tenure and ownership were intricately linked to the agricultural economy and the temporary nature of many tinworks did not encourage or require the development of a related settlement pattern. Instead, as tinworking shifted from one place to another it was generally much easier to travel any additional distance to work rather than overturning the equilibrium of the established settlement pattern in which landowners had a vested interest. There are of course exceptions, and in 1681, Nicholas Rowe of St Agnes, tinner, leased a plot of land on Mithian Common on condition that he enclosed the area and built a dwelling house within twelve months.

Whilst taking cognisance of the complex climatic, land tenure, ownership and other general economic and social conditions within the stannaries it is still worthwhile suggesting possible ways in which cassiterite exploitation may have influenced the overall economy. Thus for example, the continual shift of production within and between the stannaries may have affected the economy of the settlements, with viability declining or increasing proportionally to the accessibility of workable cassiterite. It may be therefore, that the history of settlement within the stannaries is bound to the fortunes of the industry and at least some of the deserted settlements on Dartmoor and Foweymore may be partly the consequence of a shift in production from this stannary in the fourteenth century. This explanation would certainly not contradict the excavated evidence from Trewortha and Bunning's Park, where pottery suggested that these particular settlements had been abandoned at this time.

A secondary effect of mineral exploitation on the local economy, which is known to have manifested itself in the settlement pattern is that the increased population of tinners would have raised the demand for foodstuffs, and this in turn must have influenced the price. The farmers in the stannaries responded by bringing marginal land into use to satisfy the increased demand. Thus on Dartmoor, Fox has attributed at least part of the expansion of rents and reclaimed acres to increased demands resulting from the tin industry. It also follows that a depression in the industry would have had a knock-on effect. In some instances, the resulting fall in demand could have been responsible for some of the desertions noted on the periphery of the main tin-producing areas.

Tinworks and tinbounds

The names of at least 2,341 tinworks and tinbounds are known, although the quantity and quality of information relating to each varies considerably. At one end of the spectrum all we have is the name, whilst at other sites, we have for example details of production, names of the workforce and location of the site. Sadly, the information is grossly biased, with over 97 per cent of the known sites belonging to the sixteenth and seventeenth centuries. The primary reason for this is that prior to 1496 tinworks were not registered at the stannary courts and consequently unless there was some form of dispute or other legal matter regarding a particular site they were generally not recorded. The second reason is that much of the earlier documentation has been lost, or may be held in collections that have not yet been examined. The information explosion of the sixteenth century is a consequence of the registering of tinwork bounds with the Stannary Courts, the listing of tinworks by some of the large-scale owners and more detailed accounts relating to the collection of Toll Tin. Thus for example: the Foweymore Stannary Court Rolls for 1506 refer to 12 tinworks; Peter Edgcumbe's list of tinworks in 1593 contained 57 names and in 1566 a single book contained the names of 28 tinworks for which Toll Tin was collected by Mr Pers Eggecombe. In the sixteenth century the stannary with the largest number of recorded tinworks and bounds is Blackmore with 178 (**16**). By complete contrast the majority of the tin produced at this time came from the stannaries of Penwith and Kerrier and Tywarnhail. The possible reasons for this contradiction are firstly that the tinworks of the West were more productive, and consequently a large amount of tin came from a relatively small number of mines. However, the more likely interpretation is that the documentation is very biased, with the eastern stannaries being exceptionally well represented in the sixteenth-century sources examined. This is particularly the case for Blackmore where 136 of the tinworks are recorded in the Edgecombe documentation. The existence of this single source has thus clearly emphasised one particular stannary at the expense of the others, and the failure of similar documents to survive for other areas must have influenced the distribution.

This problem of differential survival of documents manifests itself more clearly in the seventeenth-century distribution map, for in this the small stannary of Tywarnhail is represented as having many more tinworks and bounds than the combined Cornish total. The reason for this is two documents (C.R.O. DD.EN. 159 and 1588/3) which are substantial lists of Tywarnhail bounds belonging to Hugh Tonkin, Stephen Hickes, Thomas Cossan and others. Similar documents relating to the other stannaries were not found and the result is a distribution that is largely one of document availability rather than tinwork/bound distribution. This situation makes it impossible to comment on the distribution of tinworks as revealed by the site-specific documentation. The only conclusion which may be drawn from the information is that a further substantial number of tinworks must have once existed, the documentation for which has either been destroyed (e.g. the stannary records of the Hoblyn's of Nanswhyden perished in a fire of 1803) or are still to be located. It is thus unlikely that the precise number of tinworks functioning in each of the stannaries in the different centuries will ever be ascertained and the numbers represented in Figure 17 must be considered as the bare minimum.

16 Large numbers of documented tinworks and tinbounds are known from sixteenth- and seventeenth-century documentary sources. The substantial growth in the Tywarnhail stannary is of special interest, but could this simply represent an anomaly in documentation survival or availability?

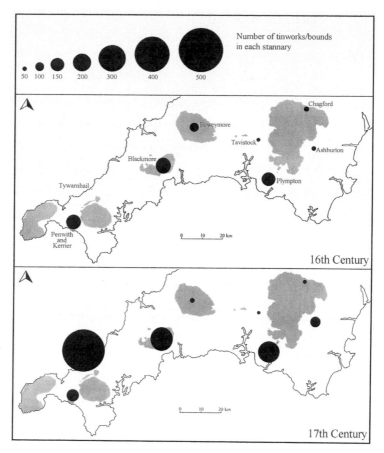

It is, however, possible to establish the relative importance of each stannary, throughout the period using the coinage figures and presenting them using a distribution map (**18**). For the purposes of this map the output from the four Devon stannaries are grouped together. In the early fourteenth century the most important Coinage town was Lostwithiel but by 1577 the picture was very different with the western stannaries now dominating the industry. This trend from east to west continued and by 1607 the great bulk of tin was coming from the west. The shift of the industry westwards is further confirmed by examining the Tribulage figures for the two stannaries of Blackmore and Penwith and Kerrier. Tribulage was a form of capitation tax paid by every labouring tinner in these two stannaries and was levied at the rate of one halfpenny per head until the Black Death (1349), and thereafter it was increased to twopence in Penwith and Kerrier, but remained at the original level in Blackmore (presumably to encourage tinners to remain in the now poorer eastern tinworks). The amount of money raised by this tax is known for thirty-five years between the late thirteenth and sixteenth centuries, and consequently the number of tinners active in these two stannaries can be calculated. In 1297 the number of tinners is similar, although there were slightly more in Penwith and Kerrier. This is in direct contradiction with the Coinage figures which indicate that there was more tin being produced on Blackmore than in Penwith and Kerrier. One possible explanation of this

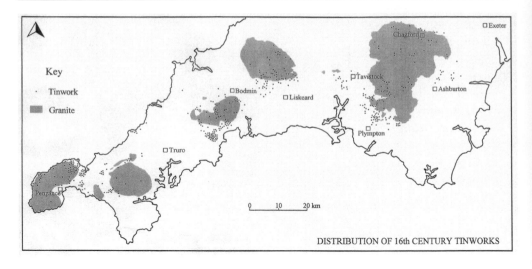

Key

· Tinwork

Granite

Exeter
Chagford
Tavistock
Bodmin
Liskeard
Ashburton
Plympton
Truro
Penzance

0   10   20 km

DISTRIBUTION OF 16th CENTURY TINWORKS

17 This map represents our current understanding of the distribution of tinworks in the sixteenth century. As research continues this picture is likely to be considerably enhanced. Marked clusters of tinworks highlight areas which were either particularly productive during this period or where documentation has survived

situation is that more tinworks in the western stannary were exploiting the lode material and were consequently more labour intensive. If this hypothesis is accepted this would date the introduction of medieval mining to sometime probably in the thirteenth century. On the basis of the evidence, this possibility should not be ignored, but it is also possible that the Blackmore tinners were more efficient at avoiding this particular tax. In the remaining three centuries the ratio of tinners between the two stannaries increased in favour of Penwith and Kerrier, although the numbers working on Blackmore continued to rise despite an ever-decreasing level of output. The amount of tin produced by each Blackmore tinner must thus have been constantly decreasing, presumably as a result of the alluvial deposits being increasingly more difficult to exploit and mines being opened up to replace those which had become exhausted.

This picture of ever decreasing production is one recorded by Hatcher for thirteenth-century Devon, where between 1288 and 1301 average production per man fell from 327lb to 145lb (148kg to 66kg). The production levels in Penwith and Kerrier were probably similar to those of Blackmore, but output continued to increase because of a proportionally larger increase in the total workforce. The most common reason given for the westward shift is the exhaustion of the eastern streamworks and the consequent opening up of the western mines and certainly both the geological character and documentary information support this contention. In the west the tin is generally held within the lodes, whereas in the east most of the mineral has been weathered and carried by fluvial action to the valley bottoms. The alluvial deposits were most easily worked and hence were extracted first, and the lodes were only turned to when the decreasing levels of production in the streamworks increased their costs and consequently made mining

18 Maps showing the weight of tin produced from different parts of the South-West at three different dates. This series of maps illustrates the shift of production westwards during the later medieval period. (Source: Maclean, 1874, 190)

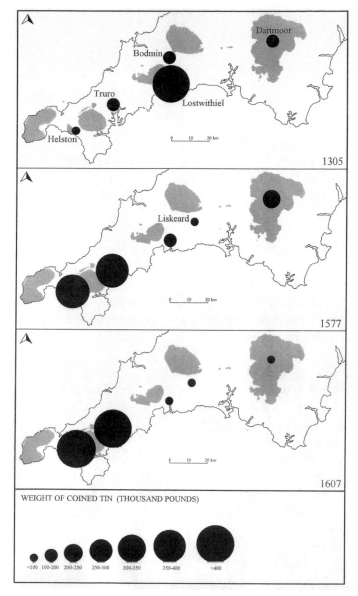

viable. The stannary documentation thus indicates that throughout the medieval and early modern period, output per person declined initially as the alluvial deposits became increasingly exhausted and then the more labour intensive mines of West Cornwall were exploited. Unfortunately, the fifteenth- to seventeenth-century mines were almost invariably reworked in the modern period by even greater numbers of miners whose impact on the works themselves and the nearby settlements has resulted in the destruction of the archaeological evidence of this important later medieval development.

5 Streamworks

Streamworks are found throughout the stannaries, although many of the best-preserved examples lie on Dartmoor and Foweymoor. Streamworks found lying beside rivers are known as alluvial streamworks because they were exploiting alluvial tin, whilst those exploiting eluvial deposits on the hillsides are called eluvial streamworks. Both types of streamwork were exploiting tin that had been detached from the parent lode and subjected to varying amounts of weathering and transport before coming to rest in the tin ground. The tinners needed to extract the tin from these deposits **(Colour Plate 8).** This was achieved using water to wash away the lighter associated clays and sands whilst leaving behind the cassiterite (tin ore) which had a higher specific gravity, and of course any large rocks. The velocity of the water needed to be controlled very carefully to ensure that as much of the worthless clay, silts and gravels as possible were removed, whilst at the same time leaving behind the valuable tin. The velocity would have been controlled by altering the volume of water and/or the gradient of the work area, which is known as a tye. A constant tye width was maintained by using banks of waste material, these have left behind diagnostic spoil heaps from which information concerning the character and development of each streamwork can be established **(Colour Plate 9).** Streaming often started at the lowest point in the valley and progressed upstream with the tye being constantly moved up-slope as the work continued. As each tye became too wide to maintain the necessary water velocity, a new one was constructed by using waste material to form a fresh lower edge. When work ceased, the final work area was left and can often be seen in the field as a narrow channel immediately next to the undisturbed valley side **(Colour Plate 10).**

Alluvial streamworking

The essence of streamworking is to separate the cassiterite from other associated minerals, and this is achieved by encouraging water to pass through the deposits. The specific gravity of cassiterite varies between 6.8 and 7.1, compared to 2.65 for quartz, 2.53 – 2.60 for feldspar and 2.94 for mica. These minerals are constituents of granite and are thus often associated with the cassiterite. The heavier tin ore thus remains behind when a flow of water removes the lighter materials. This principle was employed to separate the cassiterite from its associated 'waste'. However, the situation, in reality, is more complicated than this because the character of the deposit affects the water velocity required to remove the different constituent parts. The clay component can only be removed if the velocity is greater than 50cm/second, the silt at greater than 20cm/second, sand at greater than 10cm/second and the gravel at greater than 100cm/second. Thus, a velocity between 50 and 100cm/second would enable the clay, silt and sand to be removed, leaving behind the heavy tin ore, gravel and stones. The skill in streamworking must have been achieving the necessary velocity to remove the bulk of the waste without losing the cassiterite. The velocity would have been controlled by altering the water supply or the gradient of the tye. One would expect this need to achieve the optimum velocity, to have played an important part in determining the character of the surviving earthworks.

Contemporary documentation

There is a considerable body of documentation concerning streamworking, and from the fourteenth to seventeenth centuries there are many references to streamworks, although none are sufficiently detailed to ascertain the precise techniques employed. However, the documentation viewed together with the archaeological evidence provides an insight into the character of early streamworking. One of the earliest known surviving references to a streamwork is found within the White Book of Cornwall and is dated to 1357. This deals with the imprisonment of Abraham le Tynnere who was charged with having caused 'damage to the prince and haven of Fowey'. Abraham owned seven tinworks within Foweymore Stannary, four of which (Brodhok, Tremorwode, Greyiscome and Smalescombe) were described as 'stremwerks'. The damage referred to must have been silting caused by the discharge of waste from these works. This is confirmed by the court rolls of Blackmore Stannary for 1356, which indicate that the very future of Lostwithiel port was threatened by the discharge of refuse from the Foweymore tinworks. In an attempt to save Lostwithiel's port status, the Duke of Cornwall forbade further tinworking in the area. It was presumably as a consequence of failing to obey these instructions that Abraham was imprisoned. Streamworks were thus responsible for releasing large quantities of waste into their rivers and it is worth noting that these early attempts to prevent this activity failed entirely as time and time again, statutes were passed to try to control the practice. The earliest documentation is thus concerned with the removal, by streamers, of the alluvial soils, but as long as working of these deposits continued, pollution was inevitable as the basic purpose of streaming was to hydraulically separate the heavy cassiterite from the lighter sands and silts, and these wastes could only be removed from the streamwork in suspension. It is very noteworthy that the consequences of this industrial pollution were being felt by the fourteenth century.

Another early reference (1387), concerns a 16-year-long legal dispute between a group of streamers (working a streamwork called Coweswerk in the Lamorna Valley, St Buryan) and the landowner John Treweoff of Trove. In this case, the landowner took the law into his own hands, forcibly removed the streamers and damaged their tinwork by 'flooding the trench by which they (the streamers) had diverted the river, so that their tin was lost'. Modern documentation indicates that the probable reason for this diversion was to site the river in an elevated position relative to the tin ground, so that it could be harnessed to achieve the necessary velocity to remove the waste. Diversion trenches (leats) are found in the vicinity of all streamworks, with particularly well-preserved examples at Lydford Woods (**22**), Colliford (**11**) and Stanlake (**23**).

Much of the surviving documentation relating to individual extraction operations makes no specific reference to the type of tinwork involved. Instead, they are referred to under the broad heading of 'opus stannar', tinwork or tin bound. In a few instances the documents do indicate the type, with for example, Kerla Myne and South Trekeve Myne being referred to in Stannary Court Rolls of 1513 as 'both beme + streme', whilst the Registers of the Black Prince, in 1357, note the existence of 'stremworks' at Greyiscome and Smalescombe. In other instances, it is the name of the tinwork which strongly suggests that it was a streamwork. Thus for example, in Cornwall: Goodfortune alias The Streamwork in Lanivet; Clennacombe Streme in Linkinhorne and Gallidnowe New Streame bounds in Penwith and Kerrier, are all probably streamworks. In Devon the situation is slightly different, with rivers often being known as lakes and therefore it is probable that many of the tinworks with the element 'lake' in their name were streamworks. Indeed, confirmation for this assumption comes from Foweymore, where two early sixteenth-century tinworks were described as 'Lake Workes'.

Lewis has suggested that in 1297 the entire Cornish tin output was from streamworks. His argument is based on the difference in quality between mine and stream tin, with the former being poorer, and consequently more of this type was needed to produce a given weight of metal. By comparing the conversion rates for the entire Cornish output in 1297 with the rate recorded by Beare in 1556 for alluvial tin only, he has claimed that since the figures are almost identical all the late thirteenth century output must have been alluvial. However, there are strong grounds for doubting Lewis's argument. The measures by which black tin was sold differed between the stannaries and thus it is not possible to compare measures, as Lewis did, which refer to the county as a whole (those of 1297) with those which refer only to Blackmore (Beare). Thus, there is insufficient evidence to indicate that the late thirteenth-century tin output was solely from streamworks. The date at which mining of cassiterite first occurred is considered elsewhere, but it is important to note at this point that with the depletion of the richer alluvial deposits, the tinners must have turned their attentions to the more promising lodes. Thus, increasingly less tin was extracted from alluvial contexts, whilst the output from mines continued to rise.

Although streaming became progressively less important, at no time in this period did it cease, and certainly prior to 1700 it probably maintained substantial levels of production since deep mining was still relatively expensive and limited by the available technology. For the sixteenth century, contemporary descriptions of streaming are available in the writings of Camden, Norden (1584) and Beare (1586). Camden writing in about 1540 stated that they 'lieth in lower grounds, when by trenching they follow the Veins of Tin, and turn aside now and then the streams of water coming in their way'. These comments suggest that Camden was not fully aware of the streaming process, since he seems to be under the impression that the rivers were confined to the valley sides solely to keep water away from the works rather than to serve them. John Norden, may have understood the subject less, as he did not attempt to describe them and instead merely recorded their location as 'in the brookes, in valleys among the hills', and the streamworks themselves as 'shallowe and more easie'. By contrast, Thomas Beare, was probably so familiar with the streaming techniques, and assumed that everyone else was, that he did not give an account of them, but did introduce terminology which suggests that there was more than one type of alluvial work. Thus when considering tinworks in the low valleys he wrote that they 'are the very same workes that we call streamworks, hatch-workes and moorworks'.

Equating these names with the earthworks of surviving alluvial tinworks is difficult, but the most plausible explanation appears to be that the streamworks were the alluvial tinworks and the moorworks were eluvial streamworks. The hatch-workes on the other hand may be streamworks which were exploited using a shaft or large pit to reach the underlying tin deposit. This technique is well documented for the nineteenth century at Restronguet Creek, where an iron-lined shaft was sunk into the seabed and the tin ground extracted using traditional underground mining techniques. More relevant, however, is the timber-lined shaft at Pentewan, which had been sunk onto the alluvial tin ground (p15-16). Shafts and pits identified within many surviving streamworks should possibly be considered as hatchworks (**11**).

Modern documentation

The major drawback in turning to post-seventeenth-century descriptions of streaming techniques is that they are not contemporary with the period being examined and may not, therefore, be directly relevant. However, since the basic principle of streaming, from earliest times, remained constant, it is worthwhile examining the more recent accounts to ascertain whether they can perhaps explain the surviving earlier earthwork evidence.

The most detailed description of streamworking techniques is found within Hitchens and Drew's History of Cornwall, and such is the importance and clarity of this account, that it is given here in full:

> a stream of water is conducted on the surface to that spot where he intends to begin his operations. A level is also brought home to the spot from below, as deep as the ground will permit, and the workings require, to carry off the sand and water. The ground is then opened at the extremity nearest the sea, or the discharge of the water; from which place the streamers (for by this name these tinners are known, to distinguish them from miners) proceed towards the hill. On the ground which is laid open, the stream of water is turned in from the surface, which, running over an almost perpendicular descent, washes off the lighter parts of such ground as had been previously broken by picks, carrying them through the under level, which is called the tye, and leaving behind the sandy ore, and such stones as are too heavy to be thus removed. In this stream the men, provided with boots for the purpose, continue to stand, keeping the sand and gravel at the bottom in motion. From it they select the larger rubbish, throwing it on one side, picking from their shovels such shode as appears. The precipice over which the water runs is called the breast; the rubbish thrown away is called stent; the sand, including tin, is called gard; the walls on each side of the tye are called stiling; and the more worthless parts which are driven away by the stream are called tailings. In this manner they continue to dig or break their ground until the whole is exhausted, which is sometimes the work of many years (Hitchens and Drew, 1824, 603-4)

The different features of streamworking recorded within this account have been found during field survey and no part would be inconsistent with the contemporary descriptions previously considered. This description, furthermore, explains the parallel character of many dumps because the 'tye' (work area) would need to maintain its width if it was to continue functioning and the waste (stent) would have been utilised to form the necessary lower stiling. Thus as the work progressed up-slope, new stiling would need to have been constructed, and the material used for this operation was the stent. The shape of the surviving dumps thus reflects the original shapes of the tyes.

A further useful account of streamworking was provided by William Pryce, who in 1778 wrote:

> the Streamer carries off what he calls the Overburden, Viz. the loose earth, rubble, or stone, which covers the Stream, so far and so large, as he can manage with conveniency to his employment. If in the progress of his working he is hindered, he teems (or lades) it out, with a scoop, or discharges it by a hand pump: but if those simple methods are insufficient, he erects a rag and chain pump so called; or if a rivulet of water is to be rented cheaply at grass, he erects a water wheel with ballance bobs, and thereby keeps his workings clear from superfluous water, by discharging it into his level: meanwhile his men are digging up the Stream Tin, and washing it at the same time, by casting every shovel full of it, as it rises, into a Tye, which is an inclined plane of boards for the water to run off, about four feet wide, four high, and nine feet long, in which, with shovels, they turn it over and over again under a cascade of water that washes through it, and separates the waste from the Tin, till it becomes one half Tin (Pryce, 1778, 133)

This technique is different to the one described by Hitchens and Drew, with the overburden and tinground being removed from the tinwork by hand. This would have necessarily meant that large amounts of spoil would have been moved and then dumped, and one would expect such features to enter the archaeological record. A more revealing account of this practise was given by J.H. Collins, who wrote:

> With tin-gravels in particular, the tin-ground or pay-gravel, is often buried beneath many feet of sand, gravel, peat, or other substance containing nothing of value; and this must be removed before the pay-gravel can be dealt with. The most advantageous mode of working is to clear the over-burden away completely from a good sized space, remove the ore ground for subsequent treatment, or treat it on the spot, and then to fill up the space with the over-burden from the next section. In this manner no part of the over-burden has to be removed to a great distance, nor yet to be lifted to any considerable height. (Collins, 1875, 33)

This mode of working can be identified with the 'cuesta'-shaped earthworks identified at Minzies Down (**19**) and Colliford, where large dumps had apparently been placed in previously worked areas. The second, major point raised by Pryce was the use of pumps to drain streamworks, and this has been confirmed archaeologically by the discovery of a

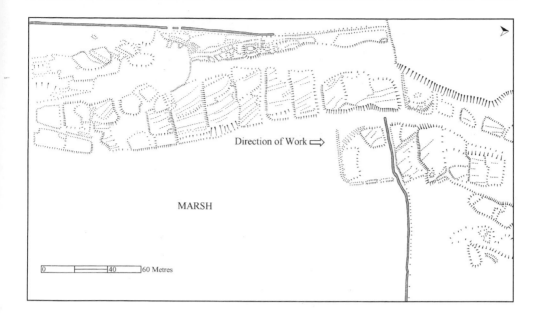

Direction of Work ⇒

MARSH

0   40   60 Metres

19 The cuesta-shaped earthworks forming part of the alluvial streamwork at Minzies Down were clearly formed by wheelbarrows. The dashed lines represent narrow gullies, each terminating at the top of each dump's scarp. These gullies are considered to be wheelbarrow ruts

'primitive pumping machine' at Nancothan, twelve feet below the surface and at Colliford, an embanked leat (*see* **11**) associated with a hatchwork may have served a wheel used for draining the operation. These texts are of a relatively modern date, but the earthwork remains found throughout the river valleys are consistent with the techniques as described. The method referred to by Hitchens and Drew, for example, would have produced the 'parallel' and 'retained' dump types recognised from field survey, whilst the techniques in Pryce and Collins' work would have produced the 'cuesta' type. In addition, the contemporary accounts, although not as detailed, being in no way contradictory, imply that the streaming methods of the eighteenth and nineteenth centuries can be considered as comparable to those used by the earlier streamers.

Absolute dating of the earthwork remains has not yet been attempted in an organised manner. From survey work alone it is possible to ascertain the relative date of some separate areas of tinworking, by examining the visible stratigraphic relationships. The potential for absolute dating does, however, exist. When a particular streamworking operation ceased, many of the tyes and pits remained open, filled with water and slowly silted up, thus preserving a complete pollen profile from the date at which the site was abandoned. Strategically positioned cores from these disused works could thus indicate the date at which any part of a streamwork ceased operations, by either having the sample C14 dated or comparing the pollen profile with other accurately calibrated examples. At Colliford a single core, from within a 'parallel' streamwork suggested that this particular example may have been abandoned around AD 600-700.

Archaeological Evidence

As a result of detailed examination of the alluvial streamworks at Colliford and Minzies Downs in Cornwall together with those at Lydford Woods, Hart Tor and Stanlake in Devon it has been possible to identify four distinctly different types of alluvial tinwork.

Hatchworks

The hatchwork technique of tin extraction essentially consisted of a small pit (the Hatch) from which cassiterite-rich material was worked and crudely dressed actually within the same pit. Water was fed into the hatch from a leat and the force of this was sufficient to remove the clay and sand particles, whilst leaving the coarser and heavier gravel, stones and cassiterite. The lighter clays were carried in suspension from the work area in a drainage level (channel), whilst the larger stones were picked out by hand and dumped at the nearest convenient spot. The remaining material would have consisted of only cassiterite and gravel which would have been dressed using temporary wooden buddles. The practice of extracting alluvial tin using hatchworks was attested by the contemporary documentation and this indicated the character of the earthworks one would expect to be associated with this type of operation. The four characteristics were: a large shaft-like hollow; an associated waste dump; a drainage level leading from the work; and a leat serving the site. Hatches are generally relatively small features and appear to represent

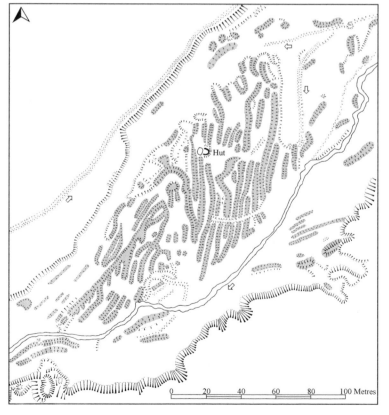

20 Part of an alluvial tin streamwork on the Hart Tor Brook. The parallel dumps are highlighted and clearly demonstrate the systematic way in which this deposit was extracted

21 *This developmental sequence illustrates the manner in which the characteristic parallel dumps found at many alluvial streamworks were formed. This illustration is based on fieldwork carried out at Lydford Woods*

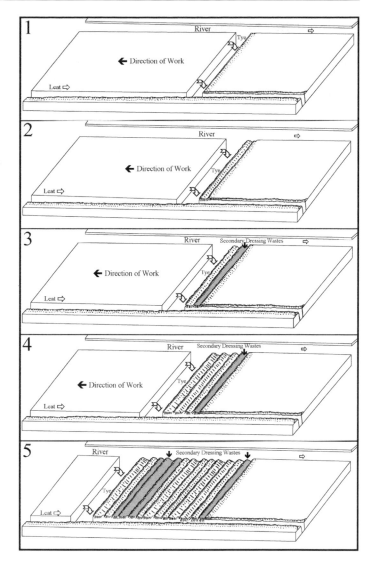

limited prospecting work within an earlier streamwork or small-scale attempts at exploitation that did not develop. Examples of hatches can be seen on the Taw River and at Lydford Woods **(22)**, where a hatch is partly cut by a later drainage level, and several previously existed at Colliford **(11)**.

Cuestaworks

A second type of alluvial streamwork consists of a series of spoil dumps with assymetrical-shaped profiles. The dip slope faces upstream while the steeper scarp faces downstream. In some instances these dumps partially overlap the one immediately downstream. The regular shape and parallel character of the waste dumps indicate that they were produced by systematic working. These dumps are very similar in character to those produced on archaeological excavations where wheelbarrows are used, and it is therefore very likely that

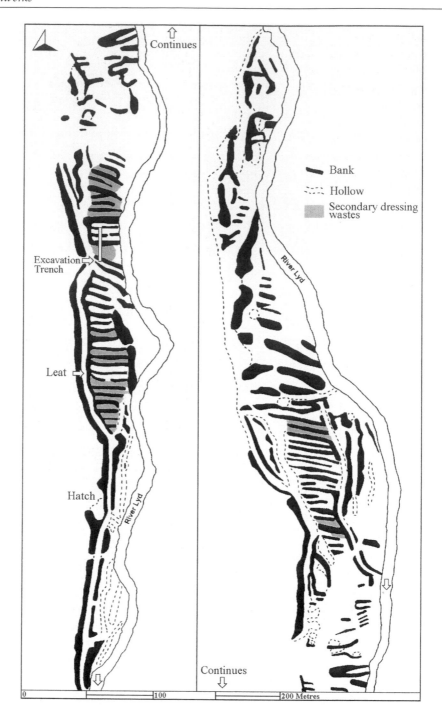

22 Simplified plan of the alluvial streamwork at Lydford Woods showing the position of the spoil dumps and areas of secondary dressing wastes. The location of the excavation trench illustrated in (24) is also shown

23 *Plans of the alluvial streamworks at Stanlake and Hart Tor Brook. The narrow dumps lying at right angles to the river are considered to represent the final phases of exploitation, whilst the wider ones lying parallel with the river are earlier and may be medieval in origin*

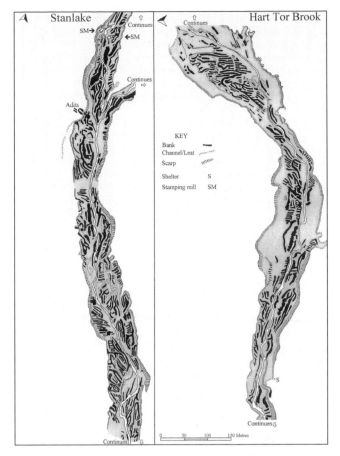

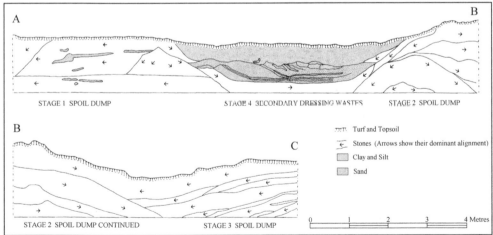

24 *Section through the alluvial tin streamwork earthworks at Lydford Woods, showing how dressing wastes had been dumped in the area between two disused dumps. The relative age of the different dumps was established and this confirmed that extraction had been carried out in an upstream direction*

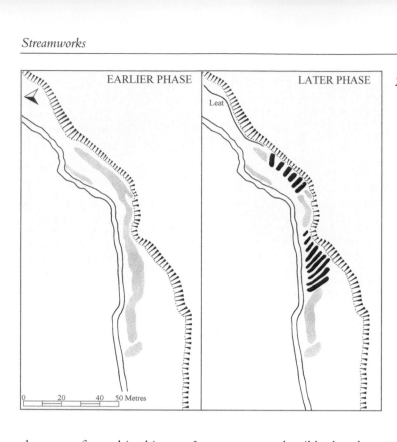

EARLIER PHASE | LATER PHASE

Leat

0   20   40 50 Metres

25 Simplified interpretative plan of an area of alluvial streamwork earthworks on the Hart Tor Brook. This plan illustrates how an early dump (shaded) was reworked at a later date. The earlier wide bank has been reworked in two places, with the resulting fresh dumps (black) being steeper, narrower and aligned differently

they were formed in this way. It seems most plausible that they were produced in areas where a considerable depth of sterile overburden had to be removed before streaming operations could start. This material was loaded into wheelbarrows and carted a short distance downstream to be dumped in an area that had recently been completed. No examples of this type of streamwork are known on Dartmoor, but a particularly fine example survives below Minzies Down on Foweymore where a series of 25 dumps are visible **(19)**. Earthworks of this type would have been produced using the streamworking method described above by Pryce and Collins.

Parallelworks

The most widespread type of streamwork earthwork includes linear banks laid out in parallel lines **(20)**. The height of these features varies considerably from as little as 0.2m up to 4m (8in – 13ft). The cross-sectional character of these banks is generally symmetrical with both long sides being equally steep although on occasion the side facing upstream may be steeper. Each of these dumps denotes the lower edge of the tye at different times during streamworking **(21)**. The steep sides of the dumps clearly meant that they could not have been formed using barrows. Given the height of many, a long-handled shovel was probably employed, with the material being waste derived from streaming operations usually up-slope of each bank. Examples of this type of streamwork are abundant in both counties, and particularly good examples can be seen in Lydford Woods **(22)** the Plym and the Meavy Valley at Stanlake **(23)** on Dartmoor and Leskernick, Buttern and Carne on Foweymore.

    Two excavation trenches combined with a follow-up auguring programme at Lydford

26 The retained dumps on the Langcombe Brook face downstream. This means that in this particular streamwork, operations were carried out in a downstream direction

Woods have shed considerable light on the character of this streamwork type. This work confirmed the processes involved in the production of the earthworks, but indicated that some of the disused tye areas were reused as dumping areas for waste sands **(24, Colour Plate 11)**. The auguring programme confirmed that only a few of the derelict tyes were backfilled in this manner. The most likely explanation is that material collected from the operational tyes was carried to portable buddles resting on earlier dumps and dressed **(23)**. The wastes from this process were thrown into the disused tye and as the work progressed further from the secondary dressing area the buddles were picked up and taken to another dump nearer to the workface.

The parallel dumps generally take two forms. The first are relatively wide banks, mostly lying parallel with the valley bottom and the second are comparatively narrow dumps, often with relatively steep sides and sometimes partly revetted lying at an angle to the river. Detailed survey at Stanlake and the Hart Tor Brook has demonstrated that the small narrow dumps are more recent than the more substantial ones. In general terms, it appears that wide dumps (often with a relatively gentle profile) lying parallel to the valley bottoms are the result of early exploitation, whilst the narrower dumps (often with a relatively sharp profile and sometimes with revetment on one side) which frequently lie at an angle to the valley bottom are the consequence of more recent exploitation **(25)**. The implications of this discovery should not be underestimated since both kinds of earthwork are easy to identify and it may soon be possible, to broadly date phases of streamworking. The next stage must be to get absolute dating evidence from both types of earthwork, but without this information, it can be tentatively suggested that the wide dumps are medieval, whilst the others are post-medieval in origin.

Retained Dump Works

The fourth type of alluvial streamwork closely resembles the third, differing only in that a drystone revetment is apparent on the up-slope side. This is probably the 'stiling' described by Hitchens and Drew. These dumps would have been formed in the same way as in the third type, but with the addition of a crude revetment to prevent waste slipping into the tye. Examples of this type of earthwork are abundant in Devon and include the streamworks on the Brim Brook **(Colour Plate 12)**, at Sampford Wood, within the Erme Valley and Blacklane Brook. However, they are much rarer in Cornwall where the only known example is on West Moor. A variation on this type of tinwork exists at Langcombe **(26)** and Stony Girt on the Avon and at Stanlake in Devon where the revetment faces downstream, strongly suggesting that these tinworks were exploited in a downstream direction. The reasons for this apparent anomaly may be related to the already disturbed nature of the ground, which would have made it more difficult if not impossible in some instances to work the ground in an upstream direction. It would, for example, have been more difficult to bring water to the tye across the previously disturbed ground and instead the water could have been taken much more easily from the nearby river.

Within many alluvial streamworks there are areas which do not appear to contain any of the earthworks considered above. The reasons for this vary from site to site, but in most cases it is a result of subsequent peat or silt accumulation that has masked the true picture **(27).** Earthworks undoubtedly survive in these areas, but are no longer visible. This situation, although unfortunate from the point of recording and consequently understanding the streamwork, does mean that these archaeological remains are no longer being eroded and any wooden or other organic artefacts will survive very well.

Eluvial streamworking

Eluvial deposits of cassiterite are those which have been detached from the lode and exposed to weathering and often transportation, but have not been sorted by alluvial action. They most commonly occur in dry shallow valleys lying above the larger rivers, but in some instances because of the character of the local topography, the eluvial deposits may lie immediately above their parent lode. This is the case when the lode lies in the bottom of a valley and the shoad, is therefore, confined. The character of the deposit varies considerably from that in which the cassiterite has been released completely from its ore, to examples where none of the mineral is free. Fieldwork has indicated that the majority of these deposits had been worked employing streamworking methods, and although there are no detailed contemporary descriptions, details from the Registers of the Black Prince strongly suggest that the fourteenth century streamwork at Treeures (Trerice) was of this type. In 1361, John de Treeures complained to the Black Prince that:

> fully sixty tinners have entered on his demesne and soil, which bears wheat,

27 Waste dumps were often obscured or even totally buried by later sediment accumulation. The field archaeologist should therefore always be aware that often only a partial picture is possible

barley, oats, hay and peas, and is as good and fair as any soil in Cornwaille, and have led streams of water from divers places to Treeures over part of his said demesne and soil, so that, by reason of the great current of water they have obtained and the steep slope of the land there, all the land where they come will go back to open moor, and nothing will remain of all that good land except great stones and gravel. (Hatcher, 1973, 45 : citing Registers of the Black Prince, ii)

This account is consistent with this tinwork being of the eluvial type, since several leats were cut to serve it (rather than the single example, typical of the alluvial streamworks), but more importantly the work was not on the alluvial floodplain, but rather on a steep slope, a location which is typical of eluvial streamworks. Thus by the mid-fourteenth century eluvial deposits were being exploited by streaming. The absence of contemporary descriptions of eluvial streaming methods is presumably because the streamers themselves did not greatly differentiate between the two types of extraction. It is, however, worth repeating that Thomas Beare may have been mentioning this type of streamwork when he referred to 'moorworks'.

Borlase, however, in 1758, described the character of eluvial deposits in some length:

Tin is also found disseminated on the sides of hills in single stones, which we call Shodes, (as is before observed) sometimes a furlong or more distant from their lodes, and sometimes these loose stones are found together in great numbers, making one continued course from one to ten feet deep, which we call a Stream; and when there is a good quantity of tin in it, the tinners call it,

28 Eluvial streamwork near South Carne on Foweymore. The reservoirs on the edge of the tinwork would have supplied water to the streamwork. Most of the visible dumps lie at a slight angle to the contour, but some lie directly across it. The part of the streamwork adjacent to the wall has no visible dumps. (Photograph by Steve Hartgroves, Cornwall Archaeological Unit: copyright reserved)

in the Cornish tongue, Beuheyl, or a Living Stream; that is, a course of stones impregnated with tin.' (Borlase, 1758, 161)

This account introduces the technical terms used by the tinners to describe eluvial deposits but does not involve a description of extraction methods. Archaeologically, the differences between alluvial and eluvial streamworks are that the latter often possess reservoirs and the earthworks are generally more distinct because recent peat accumulation is less developed. Most significantly, the alluvial earthworks are confined to the river valley bottoms, whilst the eluvial examples are situated elsewhere. Otherwise the visible elements are very similar. We must thus assume that the accounts of streamworking may be applied to either type.

Archaeological Evidence

Many examples of eluvial streamworks survive in the South-West and all belong to four main categories. The streaming process was identical to that in the valley bottoms, but the

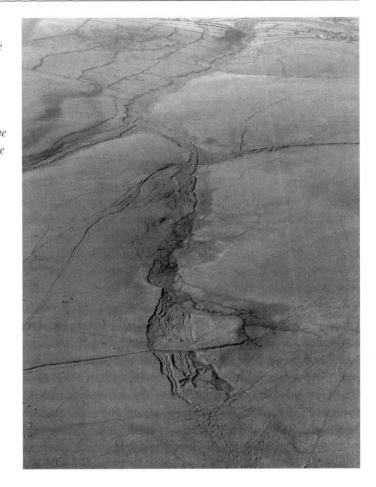

29 Eluvial streamwork on West Moor. Most of the parallel dumps within this streamwork lie across the contour. The line of prospecting pits leading upslope from the streamwork would have been excavated to ascertain whether worthwhile deposits extended uphill. The rectangular structures within the vicinity of this tinwork are peat drying platforms. (Photograph by Steve Hartgroves, Cornwall Archaeological Unit: copyright reserved)

location of the deposits meant that two factors were different. First, these streamworks were generally on steeper slopes. Second, there was insufficient water available locally, so additional supplies had to be brought from elsewhere in artificial watercourses called leats. The water was often stored in reservoirs close to the works. Leats are commonplace and can be traced for miles running along the contours of many hillsides. The reservoirs vary considerably in shape and size, but all include a bank behind which the water was held. The sophistication of water control is apparent at many sites, including St Lukes, where a single leat lies along the crest of the watershed between the St Neot and Fowey valleys and carried water to two reservoirs situated next to each other but on either side of the hillcrest — thus serving two separate streamworks in two different valleys. An even more ambitious scheme was devised to serve the streamwork on Penkestle Moor. Here, water gathered from the Dewey watershed was carried in two leats to a series of reservoirs before being channelled through seven separate and interconnecting leats to the streamwork on the southern side of Penkestle Moor (**see 13**).

The steeper slopes in eluvial streamworks have resulted in considerable variations in the layout of the diagnostic waste dumps. As with alluvial streamworking the aim of eluvial streaming was to remove the lighter wastes using water, but because of the

75

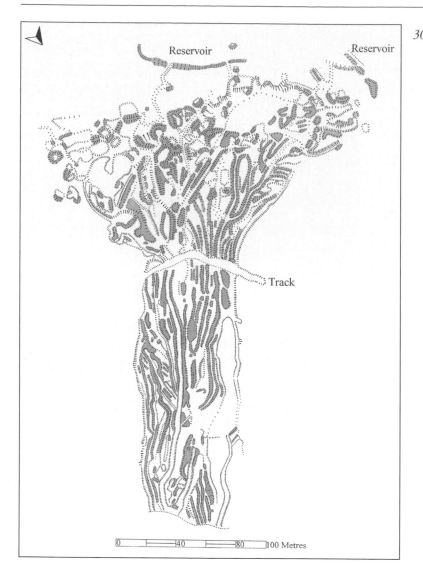

30 Eluvial streamwork at Harrowbridge (Western). Within this streamwork, situated on a relatively gentle slope at a high altitude where water was scarce, most of the parallel dumps lie directly across the contour. This form of working would have ensured that as little water as possible was required

steepness of the slope and the effect that this had on the velocity of the water, special care was needed to ensure that valuable tin was not lost at the same time. The answer to this problem appears to have been the construction of tyes at varying angles to the hillslope to achieve the necessary water velocity for streaming to successfully proceed. On particularly steep slopes the tyes were generally placed at a slight angle to the contour. Earthworks of this type are visible at a number of streamworks and include Penkestle, North (**12**), Trezelland, Carne (**28**), Redhill (**Colour Plate 13**) and Beckamoor Combe. On gentler slopes the gradient of the tye was maximised by siting the tyes directly across the contour. The streamworks at Wedlake, Raddick Hill, West Moor (**29**) and Western Harrowbridge (**30**) possess earthworks of this type. A third variation, are those sites where the dumps are curved, with the upper part lying along the contour and the lower part lying across the contour. Examples of this streamwork type are found on the upper reaches of Ivy Tor

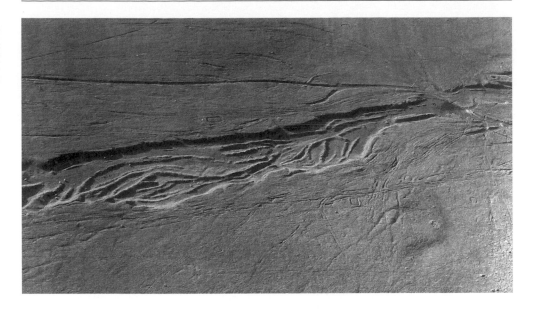

31 Eluvial streamwork at Westmoorgate. Many of the dumps within this tinwork are curved in shape. (Photograph by Steve Hartgroves, Cornwall Archaeological Unit: copyright reserved)

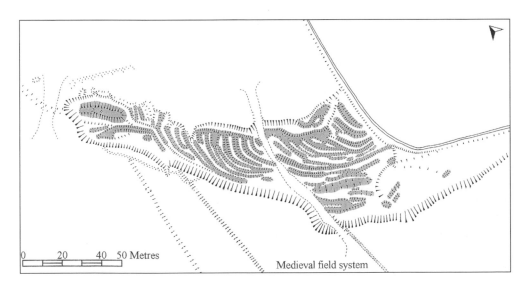

32 Eluvial streamwork at Harrowbridge (Eastern). The well preserved curving dumps associated with this streamwork lie at the lower edge of a medieval field system. This relationship suggests that there may have been a tinwork in this location during the medieval period. In 1283 there was a tinwork at "Brongelli" and this may have been at this location, although most of the surviving earthworks probably relate to later reworking

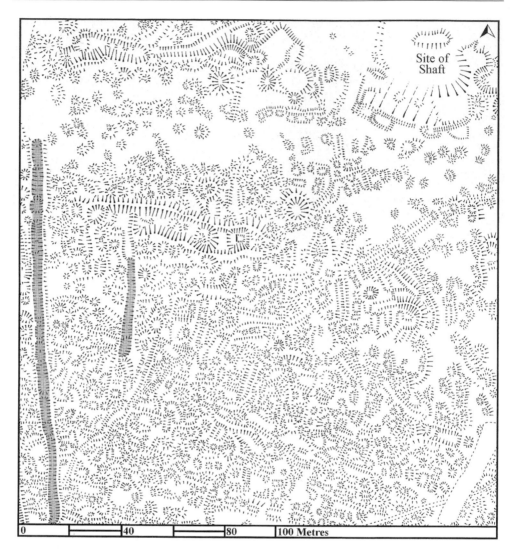

33 This small part of the tinwork on Goonzion Down illustrates the character of shoad working. The highlighted linear gullies are later prospecting trenches cut to examine the nature of the lodes in this area. The lines of larger hollows are lode-back pits

Water or Ladybrook on Dartmoor and at Rushyford Gate, Westmoorgate (**31**) and Eastern Harrowbridge, on Foweymore (**32**) and at Numphra in West Penwith. In these circumstances it is considered that the tinners were trying to maximise the flow through the lower length of the tye to prevent premature sedimentation of waste material. In some instances, the dumps are revetted on the up-slope side in an identical manner to examples found in the valley bottoms. At the streamwork on the southern slopes of Great Links Tor many of the dumps are revetted in this way with large boulders (**Colour Plate 14**).

On occasions when it was impossible to control the water velocity (perhaps, for

34 Tinwork at Kit Hill. Much of the summit of Kit Hill is pockmarked with small pits resulting from shoad working. In the foreground, dissected by the road, is an adit with associated finger shaped dump. The large gullies are openworks and the lines of larger pits are lode-back tinworks. (Photograph by Steve Hartgroves, Cornwall Archaeological Unit: copyright reserved)

example as a consequence of water scarcity), the gradient of the work area could have been altered instead and this may explain the existence of more than one type of earthwork on similar gradients within the same streamwork. At Codda, Redhill and Stanlake (**47**) different types of earthwork lie near to each other on similar gradients, and both groups may reflect periods of working when different quantities of water were available.

In some streamworks, there are areas where no waste dumps are visible. The reasons for this situation may vary. The most likely explanations include: the submergence of the dumps beneath sediments from up-slope streamworks; a paucity of waste being generated in the first place; the complete removal of dumps for 'dressing' elsewhere or at some sites the area involved may have been mined by openwork techniques. Streamworks of this type survive at St Lukes, Carne (**28**) and High Moor on Foweymore and within the Newleycombe Valley and Stanlake (**47**) on Dartmoor.

Shoad Works

A form of tin extraction which is neither streamworking nor mining is found at a few sites where eluvial deposits were exploited without the use of water. Examples of such sites are found at Goonzion, where an extensive array of pits lie next to shafts, openworks and lode-back pits (**33**), Kit Hill (**34**), Belowda and Greenburrow in Cornwall and at

79

Eylesbarrow and Hart Tor in Devon. These tinworks all lie at relatively high altitudes where large quantities of water could not be readily taken. Instead of using water to separate the cassiterite from the waste, the tin-rich material appears to have been removed from the ground by digging a series of shallow pits and taken elsewhere for dressing. This type of tinwork is referred to in the modern literature as a shoad work.

6 Mining

The three types of tinwork used to exploit lode tin were lode-back pit, openwork and shaft. These tinworks were collectively known as 'mines', since the cassiterite had to be mined from the ground.

Lode-Back Pits

Lode-back pits, are larger than prospecting pits, and survive as shallow shafts which were sunk onto the lode outcrop to extract the cassiterite encountered (**35**). These pits generally occur in linear groups following the line of the lode and each is always associated with a spoil dump. Many tin lodes have been worked 'at surface' by digging pits onto the 'backs' or surface exposures of the lode, and removing the mineral that lay above the water table. Hamilton Jenkin, in his survey of Cornwall's mines noted many examples of this type of working, which he attributed to the 'old men', a term applied to any miner who had worked for tin before living memory. There is, however, a relative paucity of reliable documentation that can be unequivocally applied to this particular form of working. The probable reason is that the miners themselves considered the technique as shaft mining, since a shallow shaft was sunk onto the lode and the cassiterite mined by the use of galleries in precisely the same manner as that of the later deep shafts. The fundamental difference is that the lode-back pits were shallow and numerous (**36**), whilst in shaft mining a lode was extracted to considerable depth by utilising very few shafts.

The slow change from lode-back to shaft mining would have occurred as technical improvements were introduced and the easy-pickings at surface were depleted. However, the basic mining techniques would have remained the same, the only difference being one of scale. It is possible that the early references to shaft mining belong more properly to this category, but because the shift from shallower to deeper working is not dated, and it was not consistent from one mine to the next, the documentation relating to shaft working as a whole is considered elsewhere.

Lode-back tinworks survive throughout the stannaries, but their distribution is best understood on Foweymore and Dartmoor where aerial photographic work undertaken by the former RCHM(E) has identified many of the surviving sites (**37**). On Dartmoor, the largest concentration lies in the area between Walkhampton Common and Holne Moor, with many large examples surviving in the Newleycombe Valley. On Foweymore the biggest concentrations lie at Hardhead, Goonzion and Minions where a particularly fine group was studied by the Cornwall Archaeological Unit (**38**). The lode-back type of tinwork is also known to be particularly common in West Penwith where unusually large

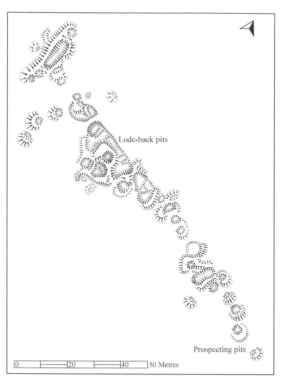

35 Part of the lode-back tinwork at Black Tor. The line of pits were excavated onto the back of a lode and the easily accessible cassiterite quarried. The smaller holes excavated along the length of the lode were prospecting pits which were dug to examine the quality of the lode. Their survival indicates that the tinners must have been disappointed with the results

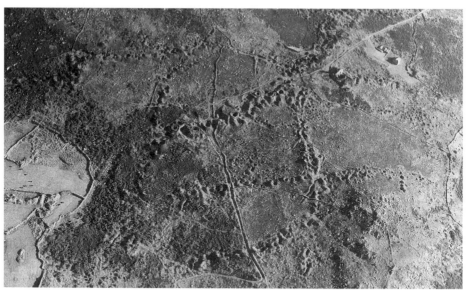

36 Lode-back tinwork on Morvah Hill in West Penwith. From the air the position of the lodes being exploited by this tinwork are clearly visible. The scale of this venture can be judged by the house standing in the upper right hand part of the photograph. Several later shafts are denoted by circular walling (Photograph by Steve Hartgroves, Cornwall Archaeological Unit: copyright reserved)

37 Distribution map showing the location of lode-back tinworks on Dartmoor and Foweymore. (Source: Devon and Cornwall Sites and Monuments Registers)

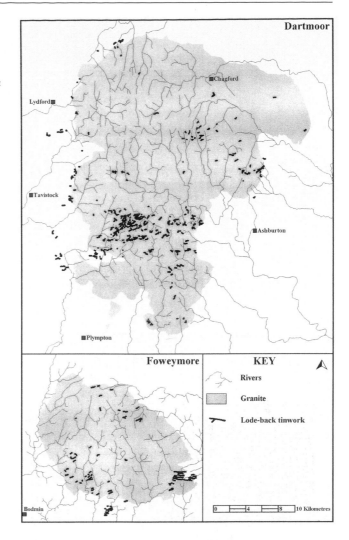

examples have been recorded at Lanyon (**39**), Bartinney and Rosewall, Caer Bran, Rosemergy, Morvah Hill (**36**), Trewey and Chun Downs.

Several lode-back pit tinworks have been recorded in detail over the years. At Noon Digery the pits were said, by locals, to have been connected by underground tunnels, a story repeated at Hobb's Hill (where they were revealed by construction work). One of the pits at this site is particularly notable because of the presence of a leat leading directly to it and continuing its course on the other side (**40**). If this leat had been merely cut by the pit one would have expected its course to have been obscured and destroyed by the dumps from this working. Instead they had been carefully placed on either side suggesting that the leat had functioned at a time when these dumps were being formed. The leat and pit are therefore contemporary and it is possible that the leat provided water to serve a wheel supported above the pit. This wheel could have been utilised to drive lifting or pumping machinery. At Penkestle (South) the lode-back pits cut into earlier eluvial streamwork earthworks confirming that streaming had been followed by mining (**41**),

38 Lode-back tinworks at South Phoenix on Foweymore. The distinctive lines of pits indicate that at least five separate lodes were exploited. On the left side of the photograph is a streamwork, whilst in the foreground are the dressing floors and buildings associated with the modern South Phoenix Mine. (Photograph by Steve Hartgroves, Cornwall Archaeological Unit: copyright reserved)

39 Lode-back tinwork and field system at Lanyon in West Penwith. The lode-back tinwork slices through earlier field systems of prehistoric and medieval date. In the valley bottom to the left of the lode-back tinwork are alluvial streamwork earthworks. (Photograph by Steve Hartgroves, Cornwall Archaeological Unit: copyright reserved)

40 *The lode-back tinwork at Hobb's Hill was served by a leat which led directly to the largest pit. A very fine group of prospecting pits ascending the hill to the south of the tinwork were probably responsible for the discovery of the lode*

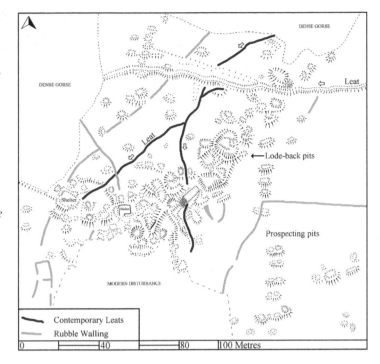

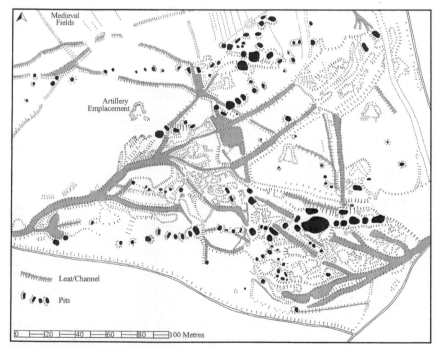

41 *The eluvial streamwork and lode-back workings on Southern Penkestle Moor. The leats and other channels are highlighted to illustrate the complex hydrological character of the workings. The lode-back pits cut the streamwork earthworks indicating that mining followed streaming*

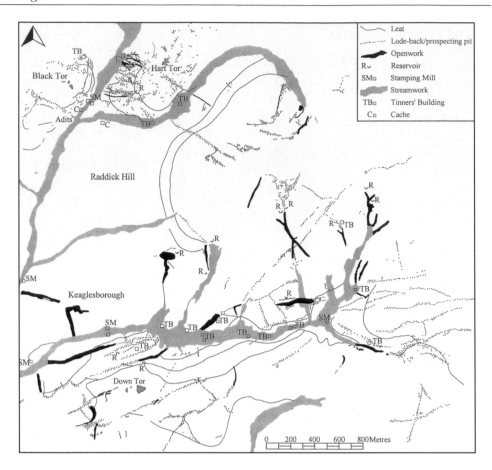

42 The archaeological remains relating to early tin exploitation in the Meavy Valley are both numerous and varied. Streamworks, mines, ancillary buildings and processing areas all exist cheek by jowl giving industrial archaeologists an opportunity to examine the relationships between the different aspects of the industry

whilst at Black Tor spoil from an adit designed to drain a lode-back tinwork, overlies earlier alluvial earthworks. At Kerrowe, the lode was exploited by pits that partially cut through the medieval strip fields, thus implying a post-medieval date for this tinwork. By contrast, work by Herring at Godolphin has convincingly demonstrated that some of the lode-back workings are medieval. One of the best recorded groups of lode-back pit tinworks survives within the Meavy Valley, where they form part of a particularly fine industrial landscape (**42**).

Openworks

Openworks are also known as beams (mainly on Dartmoor) and coffins (Cornwall) and they were formed by opencast quarrying along the length of the lode. The term openwork

43 An aerial view of the complex mining landscape at Vitifer. Substantial openworks, together with areas of streamworking, leats and shafts are all visible. Other archaeological features include at least three pillow mounds and areas of ridge and furrow. (Photograph by Frances Griffith, Devon County Council: copyright reserved)

refers to the field evidence for opencast quarrying of the lode, which produced relatively narrow and elongated gulleys, many of which still remain (**43**).

The distribution of openworks on Dartmoor and Foweymore is known from work carried out by the former RCHM(E) and whilst further work will undoubtedly refine this pattern it is clear that very few survive on Foweymore, whilst there are over 300 examples on Dartmoor, with many being found in the same areas as the lode-back tinworks (**44**).

At least 178 examples of tinworks or bounds which were either described as beamworks or had the element 'beam' in their name are known. By contrast, examples of the 'coffin' name element are more scarce, with only four in total. The small number of coffin name elements is surprising since by the eighteenth century it was the only recognised descriptive term for openworks in Cornwall, and the term 'beam' seems to have been totally supplanted. William Pryce in his Explanation of the Cornu-Technical Terms and Idioms of Tinners made no mention of beam, but wrote of Coffin:

> Old Workings which were all worked open to grass, without any Shafts, by virtue of digging and casting up the Tin-stuff from one stage of boards to another. Workings all open like an intrenchment (Pryce, 1778, 318)

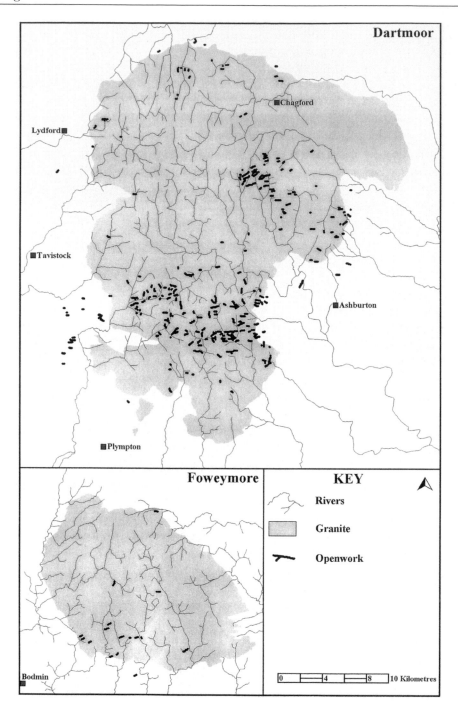

44 Distribution map showing the location of openworks on Dartmoor and Foweymore. (Source: Devon and Cornwall Sites and Monuments Registers)

Taking into account that over 300 examples are currently known of on Dartmoor alone, the combined total of 182 documented beams and coffins must be considered only a small and unknown proportion of the original number, since much of the documentation about specific sites is not detailed enough to reveal their character. For example, the tinwork at West Colliford was referred to in the sixteenth century as 'West Colyford Worke', yet fieldwork revealed a substantial openwork (**45**), a feature that could not have been appreciated from documentation alone.

No contemporary descriptions of the opencast mining technique exist and the earliest surviving account is that of Pryce, who wrote:

> they wrought a Vein from the bryle to the depth of eight or ten fathoms, all open to grass, very much like the fosse of an intrenchement . . . This fosse they call a Coffin, which they laid open several fathoms in length east and west, and raised the Tin-Stuff on Shammels, plots, or stages, six feet high from each other, till it came to grass. Those Shammels, in my apprehension, might have been of three kinds, yet all answering the same end. First, they sunk a pit one fathom in depth and two or three fathoms in length, to the east and to the west, of the middle part of the Lode discovered; then they squared out another such piece of the Lode for one or two fathoms in length as before, at the same time others were

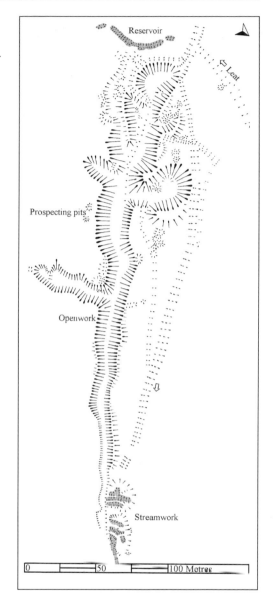

45 *The West Colliford openwork leads upslope from a small streamwork. Cassiterite from this tinwork was crushed at the nearby excavated mill*

still sinking the first or deepest ground sunk, in like manner; they next went on and opened another piece of ground each way from the top as before, while others again were still sinking in the last and in the deepest part likewise . . . Thus they continued sinking from Cast to Cast, that is, as high as a man can

46 Section across an offshoot branch of the openwork at West Colliford. Although only the upper part of this trench could be excavated, it clearly illustrates that at least some of the waste rock generated during mining operations was thrown back into previously worked areas. (David Austin: copyright reserved)

conveniently throw up the Tin-Stuff with a shovel, till they found the Lode became either too deep for hand work, too small in size, very poor in quality, or too far inclined from its underlie for their perpendicular workings. (Pryce, 1778, 141)

This comprehensive description can only be utilised to understand the techniques employed in those openworks that formed large elongated pits with vertical faces on all four sides. Many of the openworks were, however, cut into the sides of hills, thus making it possible to remove the ore downslope, rather than heaving it out of the pit (**Colour Plate 15**). If an openwork was commenced at the lowest possible point, above the water table, and then progressed into the hillside, providing that the gradient of the floor was only a few degrees above the horizontal, the work would remain free of water and the ore could be removed relatively easily through the previously extracted parts of the work. In this circumstance it would only be when the ore above the 'open adit level' was exhausted and further rich resources remained below, that the techniques described by Pryce would have been employed. Archaeological excavation of the West Colliford openwork revealed some evidence of the techniques employed by openwork miners (**46**). The tinners had removed some of the topsoil (bryle) from the lode and dumped it on either side then having mined the lode it was partly backfilled with material excavated from another part of the tinwork. It is important to note that in some of the nineteenth-century openworks,

47 *Openworks cutting through earlier eluvial streamworks at Redhill Downs and Stanlake. The openworks are highlighted with grey shading. The reservoir at Redhill Downs is unusually divided into two compartments*

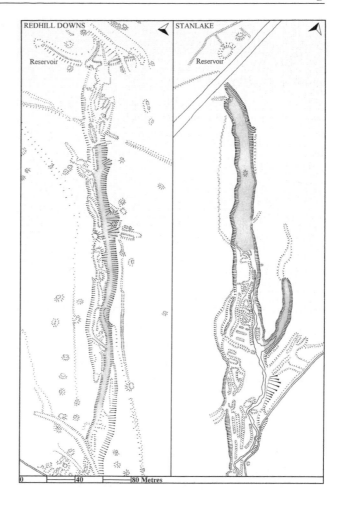

blasting, although available, was seldom used because the ore had been weathered sufficiently to allow its removal with pick and shovel alone. Thus at Minear Downs, 7-8 tons of rock were mined by each person per day and at Mulberry Hill, 5-6 tons was the average attained. Some of the early openworks may have been of this character, but others must have been extracting largely hard unweathered material, for which the rock-breaking techniques employed in shaft mining would have been utilised. It is, however, worth noting that fire setting may have been employed since the reservoirs necessary to provide the coolant have been found associated with many openworks (eg. West Colliford, Birch Tor and Newleycombe), and what is more important, being in the open air the resultant smoke would not present the same ventilation problems as it did underground.

Historians, examining the development of mining, have seen openworks as the earliest form of mine. However, caution needs to be exercised when identifying all openworks as early because it is known that many, including the examples at Minear Downs, Mulberry Hill and Rock Tin Mine, were still functioning in the latter part of the nineteenth century. The Hobb's Hill openwork is known to be more recent than the adjacent shaft and was quarried from August 1871 for a few years. It is noteworthy that the bottom of this

91

openwork was stepped to produce five distinct level terraces (or stopes according to Pryce's definition of the term), separated by vertical rock faces.

Some openworks, such as Wheal Prosper, Lanivet, even continued in use until the twentieth century. Although some openworks were either reworked or initiated in modern times, it is likely that many were indeed very early, and belong to the first recognised phase of mining. Openworks by their character could have been worked far more easily and probably more economically than shafts, and it would thus have made sense for tinners to turn to this method, once the richest and most easily accessible streamworks had been exploited. The date at which openworks first made an appearance in the stannaries is not known with any certainty, but it is likely that some were being cut before the late thirteenth century (pp 58-93). In the first instance perhaps, only the exceptionally rich lodes were exploited. The earliest of the documented openworks was 'The Beme' which was certainly in existence by 1508. It is, however, very unlikely that this represents the period at which this technique was first introduced, since the earlier documentation is both more rare and unspecific regarding the character of individual tinworks.

Some openworks probably started life as lode-back pit workings, whilst others were streamed before mining started. In some of these circumstances the earlier tinworks may have been destroyed by the openwork, though there are many examples where the lode-back workings survive in the area beyond the openwork, and others where lode-back pits survive within the bottom of the openwork (eg. Goonzion and Keaglesborough). At Down Tor, for example, a single lode has been exploited by alluvial and eluvial streamworking, lode-back pits and openworks (**42**). Many openworks are served by leats and reservoirs, although it is unclear why water was required at this type of mine. It may have been for the removal of overburden, sluicing away waste material, fire-setting, earlier streamworking and water-powered machinery. At some sites the answer is obvious, at Redhill Downs, Stanlake (**47**), Keaglesborough and Colliford (the earthworks associated with an earlier phase of streamworking are clearly visible in one small area which was not later destroyed during later openwork mining. Whilst at others, such as Hart Tor the leats were clearly excavated during the prospecting phase and there is no evidence to support the idea that water was needed for extraction purposes (**10**).

Shaft mining

Shaft mining is synonymous with underground extraction. The lodes instead of being worked only at surface are encountered and exploited at depth. The details concerning this technique are complex, but the existence of abundant, good quality, contemporary and later literature makes it possible to examine this particular aspect of the tin industry, which, because of its subterranean character is difficult and dangerous to examine in the field. The particulars relating to shaft mining and its associated terminology gleaned from various sources are most easily expressed in diagrammatic form, and (**41**) shows the character and position of the many elements in a typical mine.

Any attempt to establish the date at which shaft mining commenced is hampered by an absence of secure and positive information. There are several strong documentary indications that mining of tin ore had commenced by the end of the thirteenth century, but with the majority of this evidence it is not possible to ascertain whether the activity was limited to openworks and/or lode-back pits or even shafts. However, a 1296 reference to miners, conversant with the art of digging adits, being taken from Cornwall to work at the Combe Martin Silver mines implies strongly that shaft mining was already well established by the late thirteenth century. Further research may be able to define the origins of shaft mining more precisely.

The development of a mine in order to exploit a cassiterite lode generally took the following form:

1. Location and definition of the richness, strike and inclination of the lode.
2. Digging of a shaft to intercept the lode at the anticipated depth.
Most of the lodes were inclined sharply and thus a vertical shaft, placed a short distance to the upper side of the outcrop incline, would have generally encountered the lode. At many sites the procedure of utilising vertical shafts was not adopted and instead they were dug directly onto the lode and followed its incline downwards. This had the advantage of allowing continual recovery of 'paying' lode material instead of the barren rock encountered in the digging of vertical shafts through 'dead ground'. The primary disadvantage, however, was that the inclined character of the shaft made it more difficult to remove the ore.
3. Once a vertical shaft had reached the lode, an adit would have been cut to run along its length, to examine its character and establish whether its extraction would be worthwhile. In an inclined shaft, adits would also be cut at various depths to exploit the axial length of the lode. The material mined would either be taken to surface through the shaft or through another adit, which may have been dug from the valley bottom below the shaft, and driven to connect with the mine. The second adit would have drained the ground above this level, as well as making the removal of material easier. This type of mine is known as 'shaft and adit' and was probably the earliest form, since drainage would have presented fewer problems.

Mining of the ore would then have continued either by 'overhand' or 'underhand' stoping methods. In the former, the lode above the adit level would have been broken away whilst in the latter the ore below adit would have been extracted. As work in an 'overhand' stope proceeded the lode face would have retreated upwards, and to remain in contact with the working face the miners would have stood on the ore, which had already been detached but not removed. However, because the broken ore took up a greater volume than the lode, as work proceeded, some of the material would have been removed to the surface to maintain the necessary height. When the stope was exhausted, the ore would no longer be required to act as a platform and could then be removed to the surface, leaving a large empty cavern. By contrast, the 'underhand' stoping method required the constant removal of the ore. In this technique, mining was carried out in a downwards direction, and the working face had to be constantly cleared of all debris. After detaching some of the ore it would have been removed from the stope to reach the underlying lode. Of the two techniques, overhand stoping was the more efficient, since the ore was released

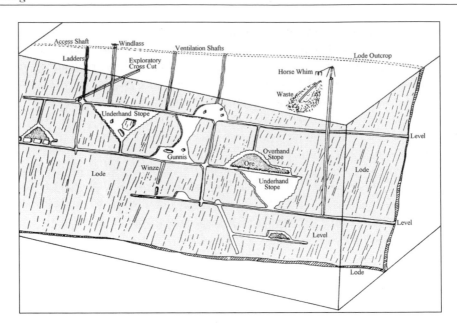

48 A simplified view of a tin mine showing the character of underground workings. (Based on Earl, 1968, 61)

to the adit level under gravity, whilst with the other, considerable effort would have been expended in continually raising the ore by hand and casting it from one step to the next to clear the working. The primary advantage of the latter technique was that ore was continually being produced, whilst in the former, the ore would only be available in large quantities once the stope had been exhausted. Mines with small capital reserves could thus have favoured the underhand stoping method to cover the ongoing expenses of labour, equipment and rent. The term 'stope' comes originally from the practise of working the lode by a series of pits in a manner identical to the method used in the openworks. This extraction technique produced a work face with a stepped profile (**48**) and each of these steps was known as a stope. More recently, the term stope has been expanded to include the whole cavity produced by mining. The use of a technique that was probably first developed above ground emphasises the similarities between opencast and underground mining, the only difference being essentially one of depth.

If the mine was to continue functioning, development work had to be undertaken. This would have taken two main forms. First, the digging of an exploratory cross cut to ascertain whether there were any other lodes in the immediate vicinity. Second, the driving of a winze (a shaft that is not directly connected to the surface) or a new shaft to investigate the known lode at a greater depth. If new economically extractable lode material was found as a result of this work, the mine could proceed until all the worthwhile ore was extracted. When new lode material was not found, the result was necessarily the closure of the mine. The most complete contemporary account of shaft mining refers not to the mines of Cornwall, but rather to those of Germany. Agricola in his *De Re Metallica* gives a comprehensive description of the techniques pertaining in the mid-sixteenth century. At this time German miners were leading the field, but it is likely that many of the techniques recorded by Agricola were employed in the British tin mines. This contention is supported by the known migration of German miners (presumably bringing their technical knowledge with them) throughout the medieval period. It is thus probably safe to assume that Agricola's account of shaft mining is a reflection of the situation then existing in Britain.

In shaft mining many obstacles that had not previously affected tinworking were encountered, and had to be overcome. The major ones were: drainage; ventilation; shoring of work places; raising the ore to surface and facilitating entry to work areas. It is Agricola who provides the largest amount of information concerning these matters, although the Cornish sources are not entirely silent.

Drainage

The most satisfactory solution to the problem of excess water in a mine was to drive an adit, from the lowest possible available point, to the mine and any water would then drain away under gravity through the open channel. Carew was certainly familiar with this practise and mentioned in his survey that:

> They call it the bringing of an adit, or audit, when they begin to trench without, and carry the same through the ground to the tin work somewhat deeper than the water doth lie, thereby to give it passage away. (Carew, 1602, 93)

The anonymous writer's observations on drainage are particularly useful and he noted:

> When we are come to any depth, and find the waters begin to annoy us, as it quickly will if any be in the work, we descend to the bottom of the Hill, where we have that conveniency, and at the lowest place begin as little a Drift (adit), as the conveniency of working or driving will permit (scarce half so big as that of the Load) on a level, till we come up to our work. (Anon., 1670, 2105-6)

The success of these adits can be judged by his comment 'if this conveniency of an Adit may be had, then our water injures us but a little'. The most significant detail to emerge from this account is that drainage adits could only be successfully used in undulating terrain where the surrounding lode bearing ground was a considerable height above the nearest river. This technique relied on gravity and consequently only those lodes situated above the river level could be drained. The digging of adits was thus one possible solution to the 'unwatering' of mines and documented examples of this procedure include: Baldew, where in 1699, the adventurers entered into an agreement with the Right Honourable Hugh Boscawen to cut an adit from the latters property; at Pitsloren in 1678, where Humfry Borlase and his coadventurers drove an adit to drain their tinwork; Trewen, where in 1684, an adit was driven by John Lanyon and finally Ball West, where in 1690 the adventurers Thomas Hawkins and Edward Coode were granted liberty by Sir William Godolphin to:

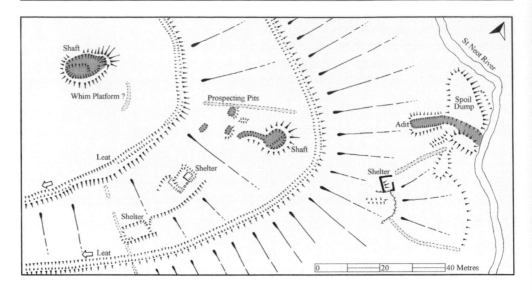

49 *Small shaft and adit mine at Trebinnick. This small mine is probably typical of the earliest tin
mines, which in most areas have been destroyed or obscured by later activity. The absence of any
processing facilities indicates that this particular mine was probably no more than a failed
prospecting venture*

> digg + worke for tyn in the Lands lyeinge betweene Baall West Bounds + the
> Tyn workes called Wheale Bargus + Wheale Bregge in Godolphin Ball + so
> carry an Auditt from the new day Shaft in or neere Wheale an Bargus to + thro
> Ball West for unwatering. (C.R.O. DD.J. 1337)

The adits at mines often survive as rectangular hollows with spoil piled up along the
long sides with further dumps lying close to the mouth. They are almost always at the
lowest point in the tinwork and therefore are generally to be found close to a stream or
river. Adits are very common in those areas containing mines and whilst many relate to
later mines, some at least must belong to early sites. The adit next to the stamping mills at
Retallack is almost certainly of an early date, whilst the one at Trebinnick, whilst undated
is of an early form. It includes a small mine with two shafts, a possible whim platform, an
associated adit and a range of shelters (**49**). At Black Tor two adits near to each other would
have provided access to and drainage from a small mine that exploited a lode which had
previously been worked by lode-back pits (*see* **9**).

The use of adits for drainage is, however, only effective for draining those parts of a
mine above adit level, and when work proceeded below this, specialised machinery had to
be introduced. Agricola, writing in Europe in the first part of the sixteenth century gives
details of the different machines then available to carry out the pumping work, and his
work is illustrated with very fine drawings of such apparatus at work. Pumping devices
available in the mid-sixteenth century included dipper wheels powered by hand, water
wheel or animals, suction pumps with hollowed out tree trunks being used for pipes and

1   *Prospecting gully cutting through a stone row at Hart Tor. Tinners were responsible for damaging some earlier archaeological remains. However, apart from the occasional robbing of a cairn, the damage appears to have been surprisingly limited, when one considers the size of some of their operations*

2   *Most tinners' buildings are rectangular in shape. This one near Leskernick is triangular, is cut into an earlier waste dump and has a fireplace in the western corner*

3 *The tinners' building at Beckamore Combe is built within an earlier streamwork. The doorway and fireplace are both clearly visible*

4 *This small circular building lying next to the Lade Hill Brook has corbelled walls and is known as a beehive hut. This particular structure lies within an alluvial streamwork and may have been used to provide shelter for a small number of tinners or their tools*

5 *Tinners' cache at Stonetor Brook. This small structure would have made an ideal place to store valuable black tin to prevent it being stolen. Originally the cache would have been completely covered over and therefore very difficult to find. Some caches remain covered and are only detectable as a small hole in the side of a dump. (Photograph by Dennis Lethbridge: copyright reserved.)*

6 *This small rectangular building hidden amongst the waste dumps at East Colliford was excavated in 1983. This work sadly provided no dating evidence, but this is hardly surprising if it was used solely to hide tin and tools*

7 *The Bunning's Park longhouse lying close to the tinworks at Colliford. Excavation of this farm-*
   stead revealed that the inhabitants were undoubtedly farmers. However, no conclusive evidence to
   support the hypothesis that they were also involved with the tin industry was found

8 *An eluvial streamwork. Systematic extraction of the tin resulted in a series of parallel dumps*
   being formed as the tinners battled to maintain the width of their tye. All eluvial streamworks
   were supplied with water from leats leading from reservoirs. (Chris Powell: copyright reserved.)

9 *Alluvial streamwork earthworks in the Plym Valley. Low light and a slightly elevated vantage point allows the field archaeologist to appreciate the parallel layout of the dumps in this part of the valley. This group of earthworks lie at right angles to the river and are considered to be amongst the more recent in this area*

10 *Eluvial streamwork at Beckamore Combe. The hollow adjacent to the scarp represents the final work area in this part of the tinwork. Material upcast during the final days lies on the left-hand side of the hollow*

11 Waste sands and silts lying between two earlier stone dumps at Lydford Woods. The use of the disused hollows between dumps for dressing waste indicates that these tinners were making efforts to reduce pollution

12 The retained dumps at Brim Brook stand over a metre high in places. This streamwork shows very clearly the systematic way in which alluvial deposits were exploited

13 Eluvial streamwork at Redhill. Within a clearly-defined gully there is a particularly fine series of waste dumps lying at a slight angle to the contour

14 The retained dumps at the Great Links Tor eluvial streamwork are composed of huge boulders

15 *On the east-facing slopes of Challacombe there are several substantial openworks. Most lie on the open moorland beyond the clearly-defined medieval field system. Some, however, have cut through the earlier fields. At least one of the openworks within the field system appears to be respected by some of the boundaries. It has been suggested that this openwork is prehistoric, however, it is much more likely that it is medieval*

16 *Mortar stone lying within the stamping mill at Norsworthy Left Bank. All four hollows were not produced at the same time. The two near hollows were formed first and then the stone shifted so that the stamp heads fell onto another part of the same stone. (Photograph by Dennis Lethbridge: copyright reserved.)*

17 *Mortar stone lying within the western room of Black Tor Falls Left Bank stamping mill. Only one face of this stone was exposed to the stamp heads. (Photograph by Dennis Lethbridge: copyright reserved)*

18 *Mortar stone at Retallack. All four sides of this stone have been used below stamping machinery with four stamp heads*

19 These mortar stones and mould stone from Little Horrabridge NGR SX 51496962 were amongst 38 mortar stones, a mould stone and crazing stone recovered from a barn when it was demolished in 1980. (Photograph by Dennis Lethbridge: copyright reserved.)

20 Lying on the surface within the stamping mill at Norsworthy Left Bank are six mortar stones, a slotted stone and another worked stone. The slotted stone probably formed part of the stamping machinery. It is likely that many of the mortar stones were originally built into the wall of the building and over the years have slumped into the mill

21 The excavated blowing house at Upper Merrivale. The furnace is denoted by the near vertical ranging rod and nearby stands the mould stone. A channel leading through the centre of the building predates the furnace and belongs to an earlier stamping mill

22 The stamping mill at West Colliford was excavated between 1979 and 1980. This excavation was the catalyst for much of the subsequent archaeological research. This building now lies below Colliford Lake

23 In the mill at Gobbett, stamping, crazing and blowing of tin were all carried out. Three mortar stones, two crazing stones and two mould stones lying within the building testify to the range of processes carried out at this site. (Photograph by Dennis Lethbridge: copyright reserved)

24 Crazing and mortar stones lying down slope from the mills at Retallack. When of no further use these stones were simply thrown out through the door of the mill and cascaded down the hill towards the river

25 Two of the buddles lying south of the mill at West Colliford. The black material is the old ground surface. On the left edge of the photograph is one of the large leats, which at an earlier date carried water further downstream to other mills

26 The mould stone lying in the blowing house at Retallack. This is the only Cornish mould stone currently known to be in its original position

27 *Furnace and mould stone at Lower Merrivale. The ranging rod denotes the right edge of the furnace. The stone immediately in front of the ranging rod is the float stone and the mould stone lies slightly nearer and to the left*

28 The float and mould stones at Lower Merrivale blowing house. The furnace is denoted by the two upright slabs on the far left of the photograph. The mould stone is filled with water and the float stone lies between the two. (Photograph by Dennis Lethbridge: copyright reserved)

29 Stamping mill at Black Tor Falls Left Bank. This two-roomed building complete with lintel across the doorway contains at least two mortar stones

finally ball and chain pumps. These were hollowed out tree trunks through which a looped chain with balls on it was pulled. Water trapped between the balls was raised and flowed clear of the pipe only once it had reached the surface. These devices could be powered by hand, water wheel or animals. Sometimes rags were used instead of balls and this variation was known as a 'rag and chain' pump.

Carew and other writers, however, were not silent on the matter of pumping machinery and although much less thorough than Agricola, they indicated that similar devices were employed in Cornwall:

> For conveying away the water they pray in aid of sundry devices, as adits, pumps, and wheels driven by a stream and interchangeably filling and emptying two buckets, with many such like, all which notwithstanding, the springs so encroach upon these inventions as in sundry places they are driven to keep men, and somewhere horses also, at work both day and night without ceasing, and in some all this will not serve the turn. For supplying such hard services they have always fresh men at hand. (Carew, 1602, 93)

The efficiency of the drainage devices is reflected in that, by the early seventeenth century Carew recorded that some mines were 40–50 fathoms (73–91m) deep, and this could only have been achieved by the utilisation of effective pumps.

Ventilation

With increasing depth, problems of ventilation would have been encountered, and again Agricola recorded the many devices employed to rectify the situation. These are: devices fitted to the top of shafts to force the prevailing wind downwards into the mine; fans operated by hand, windmills and waterwheels; bellows used to blow air directly to the part of the mine which is badly ventilated; shaking a linen cloth in the affected area to encourage fresh air to circulate and the digging of fresh shafts and adits to encourage a through draught. This method is mentioned by Carew, who noted that:

> the tinners dig a convenient depth and then pass forward underground so far as the air will yield them breathing, which, as it beginneth to fail, they sink a shaft down thither from the top to admit a renewing vent (Carew, 1602, 92-3)

Many, if not all of these devices were probably used in the Cornish mines since without them, shaft mining would have been unable to develop beyond a very primitive state. At Tremenhere, the Lanhydrock Atlas of 1696 illustrates a windmill situated within a tinwork and it is possible that this may have powered some form of ventilation machinery.

Shoring of Work Places

Mining is, and was, a dangerous occupation, and a common threat to all mines is collapse resulting in the trapping and killing of the miners. Again Agricola is the most complete source of information, and he detailed the character of shoring necessary to prevent falls.

The main point he raised is that the character of the surrounding rock determined the nature of support given to the shafts and adits. Thus:

> If the vein is hard . . . the shaft does not require much timbering, but timbers are placed at intervals, one end of each of which is fixed in a hitch cut into the rock at the hangingwall and the other fixed into a hitch cut into the footwal . . . if a vein is soft and the rock of the hanging and footwalls is weak, a close structure is necessary; for this purpose timbers are joined together in rectangular shapes and placed one after, the other without a break. (Agricola, 1556, 122–123)

Norden's, Carew's and the anonymous writer's brief accounts confirm that the same precautions were taken in Cornwall:

> Every great worke hath an overseere, and him they call captayne; whoe... provideth frames of timber to suporte the concavities where neede requireth. (Norden, 1584, 12–13)

> The loose earth is propped by frames of timber-work as they go, and yet now and then falling down, either presseth the poor workmen to death or stoppeth them from returning. (Carew, 1602, 93)

> In case the Countrey be not strong enough (as being over soaked with water from above) to support its own weight, we under-prop our Drifts with Stemples, and Wall-plates, placed much like a Carpenters square, on the one side, and over head. (Anon., 1670, 2107)

These precautions were not, however, always successful and for example at Relistian, in 1681, a rock fall killed twenty-four miners. This system of safeguarding mines from collapse has been practised in Cornish mines until the present century, and it is likely that it was adopted right from the start. Large amounts of timber would have been required, and this must have been a contributing factor, along with charcoal burning, leading to the deforestation of the stannaries.

Raising the Ore

Carew is silent on the matter of raising the ore and the anonymous writer only mentions two methods. The first was casting the ore by shovel from shamble to shamble and the second was a 'Winder with two Keebles (great buckets made like a barrel with iron hoops, placed just over the then termed Wind Hatch,) which as one comes up, the other goes down'. By contrast Agricola's account is more complete and comprehensive, with three major ways in which ore was brought to the surface being described. Amongst these are three types of windlass. The simplest type consisted of a cylinder (called a barrel) onto which rope attached to a bucket (or kibble) was wound by the tinners. This is probably the type of device described by the anonymous writer, and was probably widely used at the smaller mines. The windlass continued in use into the twentieth century, and

50 Shafts and associated whim platforms at Brown Gelly and Chagford Common. Winding machinery would have once stood in the centre of the whim platform. The animals which turned the machinery would have walked around the outer edge of the platforms. The shaft at Chagford Common is clearly more recent than the adjacent openwork because some spoil from the shaft was dumped into the already abandoned openwork

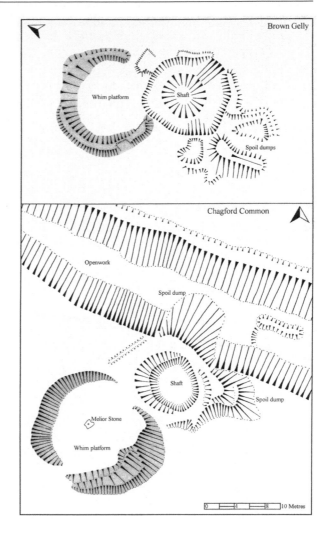

examples are known in Cornwall at a small mine near the Cheesewring and on Dartmoor at Kit Tin Mine. A variation on this theme was the wheeled windlass, which with the addition of a large wheel to the barrel of the simple windlass, enabled the barrel to maintain its inertia making it possible for the device to be operated by a single person. The third type of windlass was not given a name by Agricola but included cog wheels allowing horizontal rotary motion to be transferred to vertical rotary motion, which was used to turn a barrel similar to those of the simpler windlass types. The two primary advantages of this windlass over the two simpler varieties are, that it was less strenuous to operate and could bring up loads from greater depths.

A more efficient method of raising ore to the surface involved the use of animals as the power source. These devices are known as a horse whim and the principle of this device is that an upright barrel is revolved by the moving force of horses pushing against one or more cross beams, which protrude from the axle carrying the barrel. This action winds the rope, with a bucket on either end, onto the drum, thus allowing the ore to be raised and

51 Contemporary illustration showing four different ways of getting underground in a mine. (Agricola, 1556, 213)

the empty bucket to be lowered. The earthwork evidence for this device has survived at many sites and they must have been an important feature of the early tin industry. Whim platforms as they are known, survive as circular areas sometimes surrounded by an outer bank lying next to at least one shaft (**50**). In the centre of many whim platforms there is a large stone (melior) with a circular hole cut into it's upper face and this shows the traces of wear caused by the rotating action of the winding drum. These stones carried the vertical placed axle onto which the rope or chains wound by the action of the horse or horses. Examples of whim platforms are found throughout the stannaries and although many probably date to the modern period some must undoubtedly be earlier.

The most efficient way in which the ore could be brought to the surface involved removing it through adits, below the stopes, in carts, wagons or sleds, and brought to surface in the valley below the shafts. Most early mines probably used this technique.

These different techniques were adopted by British miners, with archaeological and documentary evidence surviving to demonstrate that at least horse whims and windlass's were employed. It is likely that some shafts, during their prolonged usage possessed more than one of the devices. Thus, for example, in the initial 'shallow' stages a simple windlass may have been employed, being replaced initially by a wheeled variety as the enterprise developed and then finally by a horse whim when full production levels were reached. This situation would, however, be very difficult to prove archaeologically, since the later devices would necessarily destroy all traces of the earlier more ephemeral forms. The final point worth considering is that since many shafts were sunk into 'dead ground' with the intention of encountering the lode at depth, they would have produced a large quantity of waste in the initial stages. Agricola noted that this situation produced 'a hillock. . . . around the shed of the windlass', and this phenomena is a characteristic of many small shafts. Examples survive at Trebinnick and Brown Gelly on Foweymore and at Hart Tor, West Vitifer and Eylesbarrow on Dartmoor.

Access to the Working Face

There are at least four main ways by which miners reached their work (**51**).

The most common method employed was probably the use of ladders, which had been fixed securely to the sides of the shaft. As the mines grew deeper the climb to and from work became greater and consequently more exhausting and dangerous. Apart from the obvious physical dangers of falling, a more substantial risk was from pulmonary disease caused by the intense exertions of climbing out of the warm mine (80°F) followed immediately by a blast of cold and wet air (30-40°F) on reaching the surface. The consequences were catastrophic to the health of the miners, and Dr Barham, writing in the early nineteenth century, noted that the instances of consumption amongst miners were two times greater than average. Mr Lanyon carrying out a similar study concluded 'that there is a great difference in the average of the longevity of the miners and agricultural labourers, in favour of the latter'. Variations on the ladder system, which may have been utilised in Cornish mines were footholes and wooden stemples. In the former, holes were cut into the side of small rounded shafts, into which the miners placed their feet, and in the second the cavities were replaced with protruding stones or timber pegs. These systems possessed all the disadvantages of the ladders and in addition were probably more dangerous.

In some instances where the shafts were vertical, the miners were lowered down in a basket. This method was similar to the technique used for raising the ore. This is the only method mentioned by Carew, implying its widespread use in the sixteenth- and seventeenth-century mines. It is, however, interesting to note that nineteenth-century Cornish miners refused to consider this method as a worthwhile alternative to the crippling ladders. This distrust may have had early origins and could suggest that the seventeenth-century baskets had a poor safety record.

If the shafts were inclined, an alternative method employed by the miners, was sliding downwards on their bottoms using a suspended rope to check the speed of descent. Ascending was accomplished by climbing using the rope to prevent them falling. The same risks of pulmonary disease and falling, would have accompanied this method.

Some inclined shafts were sufficiently gentle to enable steps to be cut, allowing the miners to walk to work. This particular approach would have been safe and desirable, but the expense involved in the digging of such a shaft could only have been borne by the most profitable of ventures, and it is unlikely that any existed. In mines with deep adits open to 'grass' many of the stopes could have been reached by a relatively level passageway, but as the ore was removed and the mine deepened, the other methods would have been employed. The problem of travelling to the workface increased with depth, and this difficulty was never satisfactorily overcome during the early phase of mining.

Mining techniques

The methods used to break the ore from the lode are considered by a number of the contemporary writers. Agricola's account is the most comprehensive, and he recorded the different ways in which rocks of varying hardness were extracted. The softer rocks were dug with the use of the pick only and the ore collected in a 'dish placed underneath to prevent any of the metal from falling to the ground'. With harder rocks there were two main techniques available to the miners. The first was to break the rock loose by blows with a hammer upon a sharpened iron tool. The miners had a variety of such instruments available to them and the precise hardness and character of the rock determined which they employed. The anonymous writer described two of the tools employed specifically by Cornish miners. These were:

> A Beele or Cornish Tubber (ie. double points) of 8 1., or 10 1. weight, sharped at both ends, well steeled and holed in the middle. It may last in a hard Countrey 1/2 year, but new pointed every fortnight at least. Gadds, or Wedges of 2 1. weight, 4 square, well steeled at the point; will last a week; 2 or 3 dayes, then sharpened. (Anon., 1670, 2104)

Carew noted that:

> Their ordinary tools are a pickaxe of iron about sixteen inches long, sharpened
> at the one end to peck, and flat-headed at the other to drive certain little iron
> wedges wherewith they cleave the rocks. They have also a broad shovel, the
> utter part of iron, the middle of timber, into which the staff is slopewise
> fastened. (Carew, 1602, 92)

The second rock-breaking technique employed was fire-setting, in which the rock was
heated by a fire kindled at the lode face. This heat was so intense that the rocks cracked,
and could then be more easily broken off from the lode. In the confined underground
workings the smoke and fumes were so dense that the mine had to be cleared of
personnel, and consequently this technique was generally employed at the end of a days
work, and only if the others had failed.

Both these techniques were not very efficient and the slowness of the task was
emphasised by Carew who wrote 'a good workman shall hardly be able to hew three feet
in the space of so many weeks'. It was not until the end of the seventeenth century that
the techniques of rock mining were revolutionised with the introduction of gunpowder.
According to tradition, blasting was initially adopted in St Agnes, but the first secure
evidence comes from the Breage Parish Register, where the death in 1689 of Thomas
Epsley who 'brought that rare invention of shooting the rocks', is recorded. The
introduction of gunpowder would have enabled the ore to be mined more efficiently thus
bringing more mines and lodes into economic production. Gunpowder helped
revolutionise mining and it heralded the change from the earlier 'primitive' form into the
Modern Genera.

7 Stamping, crushing and dressing

The black tin recovered from the streamworks and the ore quarried from the mines needed to be processed, to transform it into the valuable metal. The black tin from the streamworks was taken directly to the blowing house where it may have received a final washing prior to smelting. The ore from the mines on the other hand had to be taken first to a stamping mill where it was crushed and dressed to release the cassiterite (tin oxide) from the gangue minerals with which it was associated. Once crushed and dressed the resulting black tin was also taken to the blowing house.

Stamping

Cassiterite was separated from its ore by crushing it into small particles, and then using water to separate the lighter constituent parts from the metal oxide. In the medieval period, machinery was developed to crush the ore, and these devices are known as stamping mills, knocking/knacking mills, clash mills, crazing mills and tin mills (**52**). The first three were probably just different names for the same type of device, namely a set of water-powered trip hammers, which crushed the ore placed below them. Crazing mills, on the other hand, were very different in character, resembling grist mills. The term, tin mill, is a more general one and probably refers to any mill building in which tin was being processed.

The documentary evidence takes two different forms. First, descriptions of tin dressing processes, with the contemporary accounts of Carew, Agricola, and the anonymous writer being supported by the later work of Pryce. Second, references gleaned from many different documentary sources, relating to the individual mills that once served the industry.

The date at which stamping machinery was introduced into the Cornish mining scene is very difficult to ascertain. The earliest known reference to water-driven trip hammer mechanisms comes from China, where in AD 290 rice was hulled by this type of machinery. In Europe, the earliest examples were used in the forges of Schmidmulen in AD 1010, but by the late thirteenth century, mills for crushing ore were increasingly found all over Europe. The technique for mechanical crushing of tin ore was thus certainly available by the thirteenth century, and it is likely that stamping mills were introduced at or shortly after this time. However, the earliest documented stamping mills in Cornwall are at Penenkos, where in 1402 tin mills associated with crazing mills are recorded. On Dartmoor, we have to wait a further one hundred years for the earliest reference to a stamping mill with the mention of a mill at Ashburton in 1504.

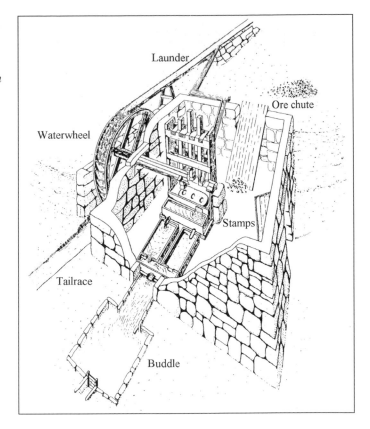

52 *Reconstruction drawing of a stamping mill showing the major components known from contemporary documentation and archaeological fieldwork (Newman, 1998, 47: copyright reserved)*

Stamping Machinery

There were two types of stamping machinery in the medieval period, namely 'dry' and 'wet'. From Carew's and Agricola's account it is clear that the only differences were that the stamp heads of dry machinery were about fifty percent larger and that no water was involved in the crushing process. The wet stamps were so named because water was passed over the ore when under the stamps, encouraging its more rapid breakdown and eventual carriage to the next stage in the process. By contrast in dry stamping the machinery had to be continually stopped to manually remove the crushed ore. In addition, the ore was usually only partly crushed by this process, and to complete the work it was often also passed through a crazing mill.

The stamping machinery consisted of a large wooden frame in which substantial, upright, iron-shod timber beams called lifters sat (**53**). The power to drive the machinery was provided by a water wheel, and by converting rotary motion into vertical movement the ore was crushed. The mill wheel axle had wooden knobs or teeth called caps attached to it, and as the wheel rotated they came into contact with tongues that protruded from, and were firmly attached to, the lower part of the wooden lifters. As the wheel continued to turn, the lifters were raised until the caps were clear of the tongues, and at this point the iron shod lifters would fall under gravity onto the ore, crushing it between the stamp heads and a hard rock called the mortar stone, on which the ore had been placed (**Colour Plate 16**). The lower, iron-shod, part of the lifters were called stamps, hence the name

105

53 Contemporary illustration showing "wet" stamping machinery being used to crush ore. A piece of bronze screen used to prevent large pieces of ore passing from the mortar box was found during the excavations at West Colliford. (*Agricola, 1556, 313*)

A—Mortar. B—Open end of mortar. C—Slab of rock. D—Iron sole plates. E—Screen. F—Launder. G—Wooden shovel. H—Settling pit. I—Iron shovel. K—Heap of material which has settled. L—Ore which requires crushing. M—Small launder.

given to the machinery. The spacing of the caps on the wheel axle determined the order in which the lifters were raised and fell, and according to Pryce, the first stamp to fall would have been the one nearest the wheelpit, which would have forced the ore under the second stamp, where it would have been crushed further before being passed onto the third, and final stamp, which fell last. From here, in wet stamps, the crushed ore was then carried in suspension through a grate at the end of the mortar box, and so on into the channel which carried it to a settling pit (**54**). The wet stamping machinery was far more efficient and therefore replaced the dry variety. This was confirmed by Carew who wrote 'Howbeit, of late times they mostly use wet stampers' and more assuredly by Agricola, who claimed to know the identity of their inventor, Sigismund Maltitz, who 'Rejecting the dry stamps . . . invented a machine which could crush the ore wet under iron-shod stamps'.

54 The earliest known South-Western illustration of a stamping mill and associated dressing floor.
(Pryce, 1778, Plate V)

The introduction of wet stamping must have had a major impact on the industry. Tin would have been crushed more quickly, effectively and cheaply and this must have reduced the overheads freeing additional finances that may have been spent on exploiting deposits that had been uneconomic before. It is likely that some mines, which had been abandoned, were brought back into production, whilst at others, the series of stamping mills serving a single tinwork could have been replaced by a single more efficient one of the wet variety. Excavations at West Colliford confirmed this, with the probable mills downstream of the excavated example, being denied the water which was instead taken to the new wet stamping mill.

Stamping mills

Documentary and archaeological sources refer to at least 223 stamping mills that were built in the period before 1700. This number must represent a minimum figure, since many were never documented, the documentation for others has been lost and large numbers have been destroyed by later activity. Figure (**55**) indicates their distribution and illustrates clearly that although many more are known in Cornwall, the majority of the surviving examples lie within the Devon stannaries. The stamping mills were primarily concerned with crushing lode-tin and thus they were necessarily situated close to the cassiterite lodes. Therefore, the distribution probably reflects the location of mining activity, a detail confirmed in Devon by Figures (**37**) and (**44**) which illustrate the distribution of known lode-back pit tinworks and openworks.

The relative absence of stamping mills on Foweymore is of interest and can be readily explained as a consequence of a scarcity of lode tin which needed stamping. The majority of the early documented mills belong to the seventeenth century. This is probably a reflection of improved documentation and the increase in mining at the expense of streamworking. Sadly, the bulk of the evidence only records the existence of a site and there are few or no details of its character. There are, however, a few welcome exceptions,

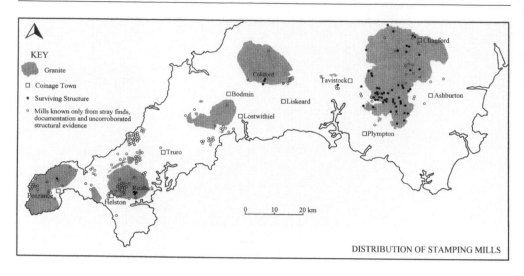

DISTRIBUTION OF STAMPING MILLS

55 Distribution of stamping mills. This map highlights the discrepancy in survival between Cornwall and Devon. Most of the mills in Cornwall no longer survive, whilst in Devon large numbers remain

and amongst these is Stephen Hickes of Truro (a gentleman with substantial interests in tinworks) obtaining a lease in 1666 from Sir James Smyth, of Chelsea, knight. The lease was to build a stamping mill at Poldice and erect buddles, frames and prepare places for laying of tin and tinstuff in a 2-acre (0.8ha) plot immediately next to the mill. Also, in 1691, the sale of the moeity of two stamping mills and a pair of associated tinwork bounds to Thomas Carlyon of St Austell, records the existence of 'mill places, buddles, races, waters and leats'.

The most comprehensive insight into the character of an individual stamping mill is provided by a series of accounts from the mill at Harvenna. A complete copy of the accounts is included below because they are a particularly valuable source of information regarding many aspects of stamping:

An account given by the Tyners of Harvenna of their disbursments about the Stamping mill.

	£ : s : d
pd. Mr Cruse for 2 new Stamp heads weightd 1Cwt : 0lb : 19 oz att 22 d.p. lb. as apears by his noate	01 : 5 : 09
for fetching the same from St Collomb + puting them in the mill	00 : 1 : 06
for 2 new Sharges (seives)	00 : 6 : 06
for a new Shovell	00 : 1 : 10
for mending the grate and puting a peice on it	00 : 1 : 00
for 3 men a day about the leats	00 : 2 : 00
for 4 men a day ruding the leats and bringing home the watter	00 : 2 : 08

	£ : s : d
for one man to helpe the Carpenter about the mill	00 : 0 : 08
	02 : 1 : 11

An account of what worke have bin Stampt at the mill by Harvenna Tynners:

Stampt: first by Henry Blake : 406 Seams at 10s per c.	02 : 00 : 06
Stampt againe by Henry Blake 90 Seme at the same prise	00 : 09 : 00
	02 : 09 : 06

Stampt by John Williams.30 Seams	00 : 03 : 00

Rich Caudles acct
An account of my disbursment about Mr Rashleigh Mill

for halfe a pound of morte to grease the mill	00 : 00 : 02

Mr John Woone + his man for cuting of tymber to make lifters+ tapets + toungs for the mill	00 : 02 : 04
Sd John Woone + his man for making the lifters and worke about the mill	00 : 02 : 04
for a halfe pound of morte to grease the mill	00 : 02 : 02

pd the Smyth for makeing new rings for the lifters and mending the old ones	00 : 02 : 06
for a new stamps head bought of Mr Cruse weighted 56 1b at 20s per cwt	00 : 10 : 00
for fetching the same from St Collomb	00 : 00 : 06
for a new grate weighted 4 pounds bought of Robert Rickard St Auste ll	00 : 04 : 00
pd the Smyth for hoaling the same	00 : 01 : 00
pd John Woone for work about the mill	00 : 01 : 04
for a pound of morte to grease the mill	00 : 00 : 05
	01 : 04 : 10

The account of our stamping to Esqr Rashleigh mill The 10th day of march 1687 first we began to stamp by ye week ★★★★★★★★★★★★ wch is accounted three weeks stamping which comes to 18 shillings next we stampt by ye hundred and carried in may and june 1688 ye first time 3 schore seam next 3 hundred and ten seem more was stampt ye same year in feburary + march 1 schore and ten seame

109

Ye whole yr we Stampt by ye hundred is 5 hundred and 3 schore seem

Mr Elleis if you please to do us ye kindness to stay for ye money before Laday dayes conyge you will do us a great freind Ship

Dated ye sixteenth of decembr 1689   Phillip Thomas And Willm. Hambly

The acct of Philip Thomas, William Hamly, Robt Hamly, John Barnes, Nichas Woone, Samll Blake, Wm Gully + Richard Cundy for Stamping Tyne in St Enoder in the years 88 + 89.

	£ : s : d
March 12 88 For 3 weeks stamping att 6s per week	00 : 18: 00
For stamping 5 hundred + 3 Score seames)	
att 10s per seames)	02 : 16 : 00
Jn ° poo ll, Jn ° Carni °, For Stamping 3 hundred seames)	01 : 10 : 00

Samll Blake, Robt	more att 10s per seames)	
Hamly, Nichas Woon		05:04:00

Memorand John Blake who works in Harvena
oweth for Stamping   00 : 02 : 06'

(C.R.O. DD.R. 5000)

A number of particularly significant points emerge from these accounts, and these are worth emphasising. The weight of the stamp heads was about 56 lbs (25.4 kg) and they could be purchased locally from the nearby village. The leats serving the mill were maintained by the tinners, but because this item of expenditure only appears on one occasion it was not possible to ascertain whether this was a regular commitment. A carpenter was employed. This person probably maintained the machinery, much of which was made of wood. The cost of stamping included the hire of the mill itself at 6s per week plus an additional payment of 1s for every 10 seames of ore that were stamped. According to Pryce a seame is 'a horse load, viz. two small sacks of black Tin' and weighed about 315 lb (143 kg). Thus in March 1688 Philip Thomas et al. stamped about 1,575 cwt (80 tonnes) of ore, Jn Pooll et al stamped 844 cwt (43 tonnes), and in 1692 Henry Blake stamped 1,395 cwt (71 tonnes), whilst John Williams stamped only 84 cwt (4.2 tonnes). The machinery needed greasing, presumably to reduce friction. The lifters, tapets and toungs were made at the mill, rather than being purchased from elsewhere. The mortar box grate was pierced at the mill. This presumably allowed the tinners to choose the size of the holes which they considered most compatible with the nature of the ore being crushed. The tinners asked for credit until the next coinage (when they would have received payment for their tin). This indicates that these adventurers shared the cash-flow problems of many tinners but that credit was sometimes given is indicated by John Blake's debt of 2s 6d.

Another useful source of information regarding the running of stamping mills is the Enys business ledger of tin loans for the period 1691-98 (C.R.O. DD.EN. 1031). During 1692-3 this document records the sale of 33 stamp heads in 13 separate transactions. Significantly, on ten occasions the stamp heads were sold in groups of three. This would strongly support the archaeological evidence that the seventeenth-century Cornish stamping mills were mainly of the triple head variety. The remaining three heads were sold singly at significantly lower prices (eg. 3s 3d instead of £2 1s 7d for three), suggesting that they were much smaller than the others because the cost of stamps varied according to their weight. The implication of this difference is that there may have been small single-headed stamping mills in existence at this time. The Enys document also contains details of accounts relating to stamping mills in the St Agnes area, and the most comprehensive, is that for Moses Carter's stamps at Port Chapell. This gives details of the items of expenditure contracted by this mill and these include:

	£ : s : d
one bulling shovell	00 : 01 : 00
2 bowls	00 : 04 : 00
2 grates	00 : 03 : 00
3 stamp heads	02 : 02 : 08
3 lifters	00 : 08 : 00
1 shovel	00 : 02 : 00
2 shovell staves	00 : 00 : 04
New frames	05 : 15 : 00

Ownership of stamping mills seems to have been largely limited to the wealthy. These people often leased their mills to tinners or yeomen with, for example, The Little Stamps at Porthkellis being leased by the Enys's for a rent of 50s. per annum in 1671, whilst in 1673, Sir Joseph Tredinham leased 'a pair of stamps called Carbous Stamps in St Hilary parish' to John James, junior, yeoman. The form of lease agreement between the landowners and the tinners provides a useful insight into the legal relationship and responsibilities of the two parties, and often gives at least an impression of the characters of individual sites.

A good example of this type of lease, is that relating to the mill called 'Little Stamps' at Porkellis which was drawn up in May 1694 between Samuell Enys (landowner) and William Gundry (tinner), and its more significant conditions are worth full consideration:

> That he the said William Gundry his Executs Adins + Assignes or some or one of them shall and will keepe + constantly mainteyne ten men and boyes att worke att the Ball of Porkellis aforesd and buddles there dureing the said Tearme of five yeares. And that he and they shall and will from tyme to tyme keepe and mainteyne cotihers in the leat + leates belonging to the said stamping mill that the casualtyes which now are or which hereafter shall be laid down there on the lands of Porkellis aforesd and Lizarea may not be wasted or carryed away with the water into any other psons lands att his + theire owne

costs and charges. And that all psons that now doe or that hereafter dureing the said Tearme shall adventure in the said Ball shall stampe accordinge to the customs or usuage of the said Ball and be well and kindly used. And that he or they shall not take or cause to be taken any mony reward or other consideration for the stamping save only the leaveings. And that he and they shall be att discharge + defray the third pte of all costs and charges in cleansening riddeing maintaineing + repaireing all such flood hatches headweares banks rivers watercourses leates bridges and wayes as now doe belonge or apperteyne to the watercourses in the lands of Boswen Lizerea + Tolcarne dureing the said tearme. And that he and they shall + will from time to time during the said tearme hereby granted pay or cause to be paid unto the said Samuell Enys ... the full and just farme of all tyn which he or they shall stampe att the mill aforesd according to the conditions of the Sett of the Ball and also the sixth pte to farme of all the Tayles + leaveings as shall be brought out of the lands of any other pson or psons + stamped att the mill aforesaid. And that he and they shall + will repair susteyne uphold and mainteyne the sd stamping mill herby granted as well in wales wheeles axelltrees + upstanders as in all other things whatsoever thereunto belonging or apperteyning in goods and needfull repaire att his + their own proper costs + charges dureing the said tearme and in the end thereof leave the same well and sufficiently repaired upheld and mainteyned (the stamping heads only excepted). And that he and they shall and will permitt and suffer the sd Samuell Enys ... from tyme to tyme dureing the said tearme to take + dispose of the water ariseing in Lizerea to carry + convey the same unto the stamping mill or mills of the sd Samuell Enys on Lizerea aforesaid called the Higher Stamps Cross Hole shall be Sett att any time or times dureing the said tearme that then itt shall and may be lawfull to and for the holder + holders of the said stamps called the Higher Stamps + Cross Hole aforesaid to carrye + stampe such tyn stuffe as they shall breake or cause to be broaken in the aforesd Ball att the stamping mills last mentioned. (R.I.C. EN. 111)

A number of important points are raised by this document and should be emphasised. William Gundry employed 'ten men and boyes'. It is not clear whether these workers were involved exclusively in either mining or processing, but it is possible to ascertain the total size of Gundry's operation. This consisted of a small tinwork served by a nearby stamping mill, both of which were leased separately to Gundry by the same landowner. The stamping mill crushed only cassiterite from the leaseholders leased tinwork. Gundry was responsible for maintaining the leat serving the mill and he had to ensure that the waste from the mill did not enter the river. Gundry had to pay only a third of the maintenance of certain facilities, the cost being shared with two other mills called 'Middle Stamps' and 'South Stamps'. In addition to a yearly rent of £3, a percentage of the tin produced had also to be paid to the landowner. Gundry was responsible for maintaining the mill. The landowner reserved the right to take some of the water in the leat serving Gundry's stamps, should the need arise. This particular clause may have caused problems for

Gundry if this option was taken up at a time of drought. It was, of course, in the interests of the landowner to ensure that there was sufficient water for all his stamps, or else the financial return would have been much reduced since a percentage of his income was directly related to the output from the mills.

Whetter's examination of seventeenth-century Cornish probate inventories has confirmed that the majority of mills were owned mainly by landowners and merchants. Thus, in the second half of the century, of 23 owners identified, 15 had estates worth over £50 and only two were worth under £20. More specifically, Richard Remfry's estate valued at £1,435 5s, in 1668, included implements of husbandry, tin stamps and an estate at Tolgus. In the seventeenth century many of the landowners ensured that their own mills were used by the adventurers by granting setts only on condition that the tin ore was crushed at the lords mill. Thus, for example, in granting the setts at Porthkellis Wartha in 1626, the Enys family almost guaranteed that the adventurers would use their stamping mills by the introduction of two separate clauses:

> 3 That noeman carry away any work to dress without the licence of the Said Landlord his heires and Assignes uppon payne of forfeiture of the Said workes unto the Said Landlord

> 4 That every compa handling a quantitie of work wrought to the Vallew of one hundred Seames and the Stamps aforesd or any of them want work shall carry and Stamp the same as aforesaid at Such times as the Landlord shall appoynt uppon three dayes warninge or Shall loose the same work by them wrought soe refusinge. (C.R.O. DD.EN. 509)

The fourth clause of this sett agreement is especially interesting since it would have enabled the Enys family to maximise the use of their mills by staggering the arrival of the tin ore and this would have prevented delays at the especially busy times before the coinages. A second example of this practice, in 1628, involved John Hodge of St Austell, yeoman, who was granted a licence to search for tin on Walter Carlyon's lands in Tregrehan, agreeing 'to stamp all tin found by him at Walter Carlyons stamping mill under the West Park, as the same may be there conveniently done in a reasonable time and at an indifferent price'. Finally, in 1697, Anthony Coode was granted a sett to 'digg worke and serach for Tynn ... at Harvenna' on condition that he 'stamp and dresse all such Tynn Stuffe as shall be gotten + found in + upon the Ennises att the Stamping Mill of the sd Jonathan Rashleigh ... if the sd mill be then in condition + fitt soe to stamp the same att the usuall rate att the sd mill accustomed'.

In conclusion, the stamping mills were generally owned by the wealthier elements within the industry and the tin ore crushed was either from works on their own land or their own tinworks managed by their employees. Mining was, of course, compared to streaming, a capital intensive operation and thus it is only to be expected that those who could afford to finance a mine would also possess the necessary stamping mills. It is, perhaps, surprising therefore, that so many tinners are recorded as only having leased stamps. The most likely explanation of this situation is that the landowners, would not sell

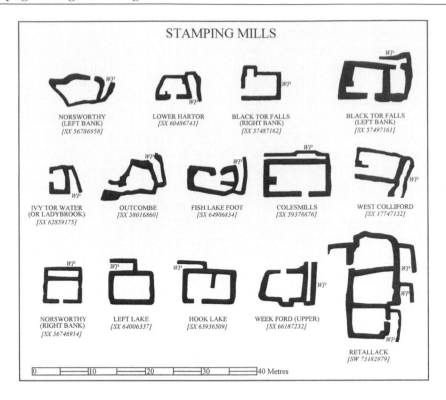

56 Simplified plans of stamping mills showing their considerably variety in shape and size

land for the building of stamps and instead preferred to lease it, as this would have ensured a constant income. The tinners, likewise, would have preferred this option, since all tinworks have a limited lifespan and the stamps with associated land would have been nearly impossible to sell at the cessation of work. In addition, by leasing, they would have done away with the necessity of making a comparatively large capital investment in land, this after all was the policy regarding the works themselves. The tinners did not purchase the land, containing the lode, but rather paid a percentage of the tin ore raised. The practise of leasing mills and plots of land for dressing purposes is thus consistent with the nature of mining.

Archaeological evidence

The archaeological evidence relating to stamping is abundant on Dartmoor, but is much more scarce in Cornwall (**56**). On Dartmoor at least 60 mills with surviving structural evidence are known, whereas in Cornwall only three remain. This discrepancy in numbers can be explained by differences in later activity in the two counties. Whilst most of the stamping mills on Dartmoor lie in locations where later mining activity and agriculture was limited, in Cornwall most of the earlier stamping mills were either

57 Distribution map showing the location of mortar stones on Dartmoor and Foweymore. The very small number on Foweymore in part reflects the lower number of lodes producing tin which needed crushing. Others may have been buried by later mining waste and some may await discovery. (Source: Devon and Cornwall County Sites and Monuments Registers)

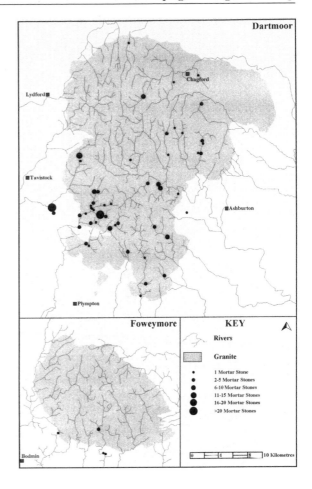

destroyed by modern mining or agricultural activities. Documentation indicates that stamping mills were originally more abundant in Cornwall, with at least 151 known compared to around 72 on Dartmoor **(55)**.

Stamping mills are very distinctive structures consisting of a rectangular building with an associated wheelpit, defined by drystone walls, often terraced into the hillslope and containing one or more mortar stones (**Colour Plates 17 & 18**). The continual pounding of the stamps onto the ore placed on these stones resulted in the formation of distinctive small saucer-shaped hollows. Some of these stones were moved within the stamping machinery on several occasions to allow fresh faces to be exposed to the pounding action and examples of stones reused in this way have been found at a number of sites. However, sometimes the stone selected was not strong enough for the task, or misuse of the machinery caused the stones to crack and break. Positive identification of stamping mills relies on finding at least one mortar stone associated with a mill building **(57)**. There are, however, occasions where no mortar stones can be found and this makes positive identification of the mill more difficult. To complicate matters further some mills with visible mortar stones may have primarily been used for smelting tin, with stamps being used to crush slag rather than ore. Many of the mills lie within, or on the edge of, alluvial

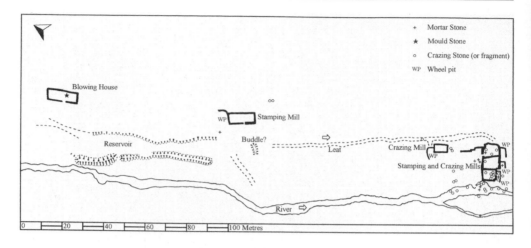

58 *The tin processing complex at Retallack is the best preserved and most complete example known. At least one blowing house together with two stamping mills and four crazing mills lie within close to each other in a manner which no longer survives elsewhere*

streamworks and survive only because they postdate the last episode of streamworking. Earlier mills in these types of locations would have been destroyed by later streaming, leaving behind only isolated mortar stones displaced from their original locations (**Colour Plate 19**). Many examples of mortar stones being found without any associated mill building probably came from mills destroyed in this way and are particularly abundant in Cornwall where of the 49 known stones, 37 are no longer associated with their mills. A further useful general observation which may be made about mortar stones, is to wonder at their location, often in the upper rubble of disused buildings and sometimes built into the walls of the building (**Colour Plate 20**). The most plausible interpretation for this phenomenon, is that many of these stones had previously been discarded and when the surviving mill was constructed they were reused in the new building. The implication is, therefore, that many of the surviving mills were rebuilt on at least one occasion during their active life.

The most extensive surviving processing site lies at Retallack, where at least six separate mills lie in a tight cluster within 240m of each other (**58**). Field survey suggests that within this complex are the remains of at least one blowing, two stamping and four crazing mills and documentation suggests that the site operated in the early part of the sixteenth century.

Excavation of stamping mills has been limited to two at Colliford on Foweymore and one at Upper Merrivale (**Colour Plate 21**) on Dartmoor. The full results of the Merrivale excavation are still awaited, but this site is atypical because although two separate stamping mills were examined, one had been converted to a blowing house. This certainly does illustrate the dangers of identifying sites from surface indications alone.

59 Aerial photograph showing the openwork, tin mill, leats, shafts and alluvial streamwork at West Colliford. The mill in the centre of the photograph is highlighted by the excavation trenches and lies at the lower end of the openwork. (David Austin: copyright reserved)

Colliford Mills *(Colour Plate 22)*

Much of the available archaeological information concerning stamping mills has been derived from the excavation of two sites at Colliford. The West Colliford mill lay at the lower end of a large overgrown openwork (**59**), which must have been the primary source of cassiterite. The structure manifested itself as a rectangular earthwork with a shallow hollow on the northern side and a mortar stone sitting vertically in the southern part of the building. Immediately to the south of the mill structure were two elongated hollows that were interpreted as buddles, this hypothesis later being confirmed by excavation. The earthworks of two leats were traced upstream for 800m and again excavation demonstrated that they had both served the mill and its buddles. Excavation revealed a rectangular drystone walled mill, with a finely preserved wheelpit, the hollow in which the stamping machinery had stood, two separate channels through which crushed ore was carried in suspension to buddles, two separate occupation levels and the many artefacts which helped date and promote understanding of the technology employed. However, much of the information regarding the site's complex history and development came from the leats above the mill. Here, sections were employed to identify seventeen separate phases of activity and, although many of these were probably a result of maintenance work, the site was deserted twice and modernised on three occasions (**60**). This suggested that tin processing had been carried out at West Colliford over a prolonged period, although the ceramic evidence by contrast was largely sixteenth and early seventeenth century and

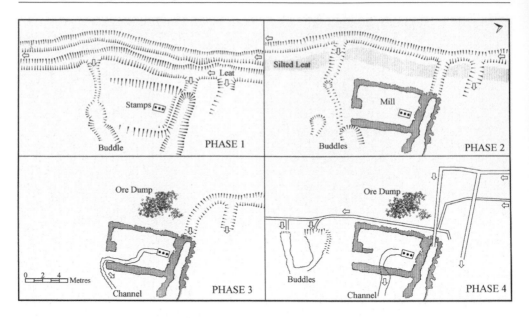

60 Excavations at West Colliford stamping mill revealed a very complex developmental sequence including at least 17 separate phases. These phases can be resolved into four major ones. During Phases 1 and 2 the ore was crushed using dry stamps. In Phase 3 wet stamping was introduced and in the fourth phase all of the water was carried to the mill in timber lined channels

supported the hypothesis that the activity was of shorter duration. However, pollen analysis of a buried soil located below the upcast from an earlier leat suggested a late fifteenth or early sixteenth century origin. The most recent pottery on the site belonged to the early seventeenth century and this combined with 1690 bounding documentation which recorded the existence of 'an old decayed Mill' indicated that the mill was already abandoned by this time. Thus the mill probably operated sporadically between the mid-fifteenth and early seventeenth centuries. The history of tin extraction is one of output fluctuations caused by political and economic factors and much documentary evidence survives to indicate that reworking or further exploitation of tinworks was a common feature. The excavated evidence from West Colliford certainly supported this general explanation, but unfortunately it was impossible to date accurately and thus ascertain the causes of each abandonment, which may have varied from national events causing falls in the price of tin, to an argument between adventurers, or perhaps a death.

A further useful consequence of the excavation was that it proved the important features necessary for accurate interpretation could only be found by this technique. Prior to excavation, the earthworks suggested that the mill was a simple structure, with a limited life (because only one mortar stone was visible) served by two leats. However, two major leats had been levelled leaving no earthwork trace, another three were timber lined and again left no clues for even the most observant fieldworker, and within the mill itself one major channel was capped and the other backfilled to obscure this important detail.

61 View of the interior of the stamping mill at West Colliford. The channel leading under the two lower ranging rods carried material in suspension from the stamps which were situated in the right hand side of the building. (David Austin: copyright reserved)

Excavation indicated that there were further works downstream requiring water, and that the wheel type had been changed from breast shot to pitch back or overshot. The large amount of useful data generated by this excavation emphasises that field survey can only give a limited, yet important, amount of information regarding a site. Only excavation can produce conclusive evidence concerning the character of the site, method of working and of course, the date. The complex nature of the site should not have been surprising since intensive industrial activity is likely to produce a complicated archaeological record. Factors such as the constant need to update the efficiency of the layout, and complete changes of ownership as one group of adventurers is replaced with others, would certainly have been conducive to continual development. Thus at West Colliford the fieldwork identification of a medieval stamping mill was confirmed by excavation (**61**). In addition it proved possible to establish that it existed in some form or other for about two hundred years, originally housing a dry stamping device which was later replaced by wet stamps. Dry stamping was not an efficient crushing method and whilst this operated at the excavated mill the leats leading downstream may have served other stamping or crazing mills, which were needed to cope with the output from the openwork. However, with the introduction of wet stamping, a single mill was able to cope with the ore output and the leats serving the other mills were backfilled and levelled. The introduction of wet stamping would have so increased the efficiency of the tinwork that they could have exploited the ore which had been previously uneconomic to extract and consequently the openwork was probably greatly extended after the introduction of this technological innovation. However, eventually the remaining ore that could be economically extracted was exhausted and the site was abandoned, remaining only as a useful landmark for late seventeenth-century adventurers' bounds.

The second mill to be excavated was at East Colliford, here a small rectangular terrace had been cut into the valley-side slope a short distance above the floodplain, and was associated with a hollow on one side and a single leat. This structure had no surviving walls, but the existence of a leat serving a wheelpit which was directly associated with a deliberately levelled terrace suggests that water-driven machinery had once operated in this area. This primitive mill was probably of an early date and no channel was found within the mill suggesting that dry crushing or grinding had been practised.

Crazing

Crazing was the process used to grind either partly crushed ore from the less efficient dry stamps or the wastes from the dressing floors. The purpose of this operation was to reduce the ore to a consistency where efficient hydraulic separation could be achieved. The most complete account of crazing is that of Agricola who noted that three types of mill were used for grinding gold and tin ore. Two of these mill types were very similar and consisted of a wooden frame in which two circular millstones were placed horizontally, one upon the other. The ore was fed in through a hole in the top stone, into the space between the two, and then ground to a fine powder by the upper stone rotating against the lower stationary one. The ground material passed out between the stones and was collected in a trough from which it was removed for further processing. The primary difference between these mills was that one type was powered by a vertical axle linked by a cog to a water wheel, whilst in the other the moving force was either men or animals. This type of machinery would have employed millstones similar to those found at sites in the stannaries. The third type was powered in a similar manner but the grinding stones were of a very different character. The lower stone had a large cup-shaped hollow in which a circular convex stone rotated. No stones of this type have been found as yet, and it is unlikely that this variation was adopted in England. The documentation relating to crazing mills, in contrast to that for stamping mills, is largely incomplete. This is possibly because in the period when crazing mills were grinding the ores from the dry stamping mills the documents generally only refer to the stamps, and not to the crazing process, which may have been considered as an integral part of the crushing procedure and therefore, not worthy of a separate mention. There are however exceptions and at Penenkos, in 1402, both the crazing mills and tin mills are mentioned, whilst at Cosgarne in Gunwalloe parish, a crazing mill and stamps are recorded in 1613. From the mid-sixteenth century many of the crazing mills were abandoned because of the introduction of wet stamping techniques which did not necessarily require a crazing facility. Crazing mills did not, however, disappear, since they were still sometimes employed to reprocess the coarser wastes from the dressing operations, but their numbers would have been reduced, and this is reflected in the comparative shortage of references.

Crazing mills

The number of known crazing mills in the South-West is surprisingly few, considering that they were once abundant. The recognized surviving sites in Devon are Gobbett (**Colour Plate 23**), Outcombe and Yellowmead, whilst at Retallack in Cornwall (**Colour Plate 24**) the remnants of at least four mills together with two complete crazing stones and 35 fragments are known. Crazing stones from Cornwall are, however, slightly more abundant with examples being found in the Loe Pool Valley, St Agnes and Godolphin.

Dressing

The tin ore, having been crushed in the mills, was carried to the nearby dressing floors where water was used to separate the lighter wastes from the heavier cassiterite (**62**). The anonymous writer's account of The Manner and Way of Dressing Tinn is the most complete contemporary description of the techniques employed. On leaving the wet stamps the tin was carried in suspension:

> into the Launder, (ie. a trench cut in the floor, 8 foot long, and 10 foot over,) stopt at the other end with a turf, so that the waters runs away, and the Orc sinks to the bottom: which when full is taken up (ie. emptied) with a Shovel.

The ore in the launder was then:

> divided into three parts, i.e. the Fore-head, the Middle, and the Tails. That Ore which lies in the Fore-head, i.e. within 11/2 foot of the grate, is the best Tin, and is taken up in an heap appart. The Middle and Tails in another, accounted the worst.
>
> 3. The latter heap is thrown out by the Trambling buddle i.e. a long square Tye of Boards, or Slate, about four foot deep, six long, and three over: wherein stands a man bare-footed with a Trambling-shovel in his hand to cast up the Ore, about an inch thick, on a long square board just before him as high as his middle, which is termed the Buddle-head, who dexterously with the one edge of his Shovel cuts and divides it long wayes in respect of himself, about half an inch a sunder; in which little cuts the water coming gently from the edge of an upper plain board carries away the filth and lighter part of the prepared Ore first, and then the Tin immediately after all falling down into the Buddle, where with his bare foot he strokes and smooths it transversly to make the surface the plainer, that the water and other heterogeneous matter may without let pass away the quicker.
>
> 4. When this Buddle grows full, we take it up; here distinguishing again the Fore-head from the Middle and Tails; which are trambled over again: But the Fore-head of this with the Fore-head of the Launder are trambled in a second Buddle (but not different from the first) in like manner: The Fore-head of this, being likewise separated from the other two parts, is carried to a third, but

62 Contemporary illustration showing a triangular shaped buddle being used to dress tin. A buddle of this shape was excavated at West Colliford (Agricola, 1556, 340)

Drawing, Buddle, whose difference from the rest is only this, that it hath no tye but only a plain sloping board, whereon'tis once more washed with the Trambling shovel, and so it new-names the Ore, Black Tin, i.e. such as is compleatly ready for the Blowing house.

5. We have another more curious way termed Sizing, that is, instead of a Drawing Buddle, we have an hairen Sieve, through which we sift, casting back the remainder in the Sieve into the Tails, and then new-tramble that Ore. After the second trambling we take that Fore-head into the second Buddle, and dilve it (i.e. by putting it into a Canvass Sieve, which holds water, and in a large Tub of water lustily shake it) so that the filth gets over the rim of the Sieve, leaving the Black Tin behind,—-.

6. The Tails of both Buddles after two or three tramblings are cast out into the first Strake, or Tye, which is a pit purposely made to receive them; and what over-small tin else may wash away in trambling.

The particularly coarse part of this material was then ground further in the crazing mill, before being dressed on a reck which was:

a frame made of boards about three foot and a half broad, and six long, which turns upon two posts, so that it hangs in an equilibrium, and may, like a Cradle, be easily removed either way, with the shovel and water, and made ready fit to be used (Anon., 1670, 2108-2111)

Carew's account, on the other hand, is much more superficial, but is worth noting in full since it enhances that of the anonymous writer.

> The stream after it hath forsaken the mill, is made to fall by certain degrees, one somewhat distant from another, upon each of which at every descent lieth a green turf, three or four foot square and one foot thick. on this the tinner layeth a certain portion of the sandy tin, and with his shovel softly tosseth the same to and fro, that through this stirring the water which runneth over it may wash away the light earth from the tin, which of a heavier substance lieth fast on the turf......After it is thus washed, they put the remnant into a wooden dish, broad flat, and round, being about two foot over and having two handles fastened at the side, by which they softly shog the same to and fro in the water between their legs as they sit over it, until whatsoever of the earthy substance that was yet left be flitted away. Some of later time, with a sleighter invention and lighter labour, do cause certain boys to stir it up and down with their feet, which worketh the same effect. (Carew, 1602, 94)

These two accounts describe the seventeenth-century tin dressing techniques in a clear and concise manner and a commentary is thus not required. However, both leave some points unresolved, and it would be useful to consider these briefly.

How efficient was the dressing process? Unfortunately, the evidence is contradictory. On the one hand the documentation suggested that a relatively large amount of tin was lost in the dressing process, and this made reworking of wastes economically viable. By contrast, mineralogical analysis of waste from the excavated buddles at West Colliford, by Richard Scrivener revealed that 'cassiterite and other heavy species are rare constituents in the fine tailings' and was completely 'lacking in the coarser grades (2mm – 125microns), from which it was presumably efficiently extracted'. The reason for this contradiction is not entirely obvious but one possibility is that the dressing workers at Colliford were unusually efficient, and reworked the tin deposits until all the cassiterite had been extracted. This particular approach would have been very costly, but presumably the tinners felt that tin recovery had to be maximised to make the operation viable. At other sites, it may have been felt that maximum profit could be achieved by extracting only that tin which could be easily recovered, and that the additional labour necessary to obtain the remainder was not worthwhile. The skill of the tinners was thus, probably not in recovering all the tin, but rather in being aware of how much work was necessary to maximise profit. Thus in those instances with an abundant supply of rich ore, only the most accessible tin would have been extracted since to spend additional time and money on obtaining the remainder would have been wasteful when the buddles could be working fresh material more profitably. The amount of tin discharged with the waste is thus likely to have been directly proportional to the quality of the lode material. At Colliford therefore, it is likely that the adventurers were mining a relatively low grade ore that had to be carefully dressed to allow the enterprise to remain profitable.

Another factor which may have affected the character and extent of tin recovery was that many tinners did not possess their own dressing floors and instead shared those belonging to the landowner. This was particularly relevant to cases where the adventurers were bound by Sett agreements. The 1626 Porthkellis Wartha Sett agreement is very explicit about the conditions under which the adventurers were permitted to use the buddles:

> That Such persons as Shall dress ye Crop of theire Tynn wrought in the Said Land Shall uppon delivery of theire Tynn remove theire tayles backwards in Some convenient place and that it shall be lawfull for others to use the Buddles to dress their Crop Tynn until the next coynage and after that the first dressing company to have the Buddles to dress theire tayles + if any of theire tayles undrest above one yeare next after the delivery of theire Cropt Tynn that then the Same to remayne at ye benefitt of the lord of the land or owner of the same mill and mills as casuallties (C.R.O. DD.EN. 509)

Under this agreement the use of the buddles was clearly divided into two separate periods. Immediately before the coinages only the best part of the tin ore could be dressed, whilst the processing of the poorer, and less rewarding parts was kept until after. These adventurers thus did not have the option to carry out the complete dressing operation as described by the anonymous writer, and they may have been less inclined to return to the buddles to rework relatively poor quality material, whilst further rich deposits lay awaiting extraction. To encourage the tinners to return to the buddles after the coinages, the agreement included the proviso that the material would revert to the landlord, if it was not reworked within a year. It is unlikely that this measure was universally successful and particularly poor quality tayles were probably often abandoned. Sett agreements of this type probably affected the character of tin recovery, since the additional interests of the landowner would have perhaps biased the tinners' perception of the profitability of further intensive reworking. A remaining problem concerns the identification of those responsible for carrying out the dressing work. Both specialised dressing floor workers and tinners with mining responsibilities were probably involved to a varying extent from one tinwork to another. There are recorded within the documentation many instances of dressing being limited to only short parts of the year. These operations were known as 'washes' and gave the adventurers, who were normally absent, the opportunity to inspect the tin as it was produced, and more importantly reduced the chances of cheating and limited the possibility of theft by ensuring that the valuable black tin was only held at the vulnerable tinworks for a short time. Washes are documented at a number of sites and these include: Lanyon in 1632, Porthkellis Wartha in 1626 and Castle Park in 1586. Greeves argues persuasively that the entire dressing process was not carried out at the washes, but instead only at the final stages of the work. A clue to the identity of the specialist dressing floor workers of the seventeenth century was provided by Carew, who noted that boys were employed, and during this period in the Trevaunance Valley 'great numbers of Boyes and humane youths are employed about washing, vanninge or cleansing tin'. The important decisions involved with dressing would have been taken by a skilled person with many years of experience, but the bulk of the work was probably carried out by boys, who in later years were perhaps largely involved in the mining. The use of boys on the dressing floors continued into the eighteenth century and Borlase noted that many as young as seven were employed 'all day long, summer and winter alike'. The first date for which details are available concerning the character of the dressing workers is 1836, when at Consolidated Mines, 335 men, 327 boys and 753 women were employed 'at surface'. Many of the men were not employed on the dressing floors and the bulk of the work was carried out by women and boys. It is interesting to speculate on the earliest date when women first made a sizeable appearance on the dressing floors. It is possible that some were employed at the early mines, though evidence is lacking at present.

Archaeological evidence

In both streamworks and mills the cassiterite had to be separated from its associated constituents, and this was achieved by taking advantage of the metal's high specific gravity, and using water to remove all less dense matter. In the streamworks the primary dressing was carried out within the 'tyes' (work areas) and only the secondary process, which produced the tin suitable for smelting, required specially constructed apparatus (buddles). Whereas within the mills all the crushed ore needed to be dressed within specially constructed pieces of equipment. From the contemporary descriptions one would have expected only the buddles set into the ground to have left an earthwork trace, with frames leaving only postholes and kieves (wooden tubs in which the tin sands were concentrated) providing no clue to their location. Thus survey will only have the potential to identify one aspect of the dressing process, and these buddles will be visible only as elongated hollows lying across the contour, possibly served by a leat and sometimes associated with the slimes which were the waste product. At West Colliford, elongated hollows south of the mill were identified from field survey as being buddles, and their excavation confirmed this interpretation (**63**). Both were gently sloping stone lined hollows each served at its upper end by a leat and accorded very clearly with those described and illustrated by contemporary observation (**Colour Plate 25**). However, these excavations also revealed that field observations alone cannot be relied on to identify all the buddles associated with a mill and a further three were found below later dumps which had hidden them completely (**64**). Dressing, by its very nature, produced a large amount of waste and during the operational life of a mill many of the buddles were likely to have become disused and then buried by dumping which would have totally hidden them from field survey. Consequently, only the buddles belonging to the final phase of activity on a mill are likely to have survived. In addition, later reworking of wastes from a mill may have resulted in the destruction of many of these buried structures. The other major feature of the dressing process at West Colliford was the slime pond where tailings from the buddles were allowed to settle and dry before being redressed to extract tin lost in the first attempts (**64**). This pond had been retained by turf and stone banks, but again none of this had been visible prior to the excavation, because of the accumulation of waste both around and over

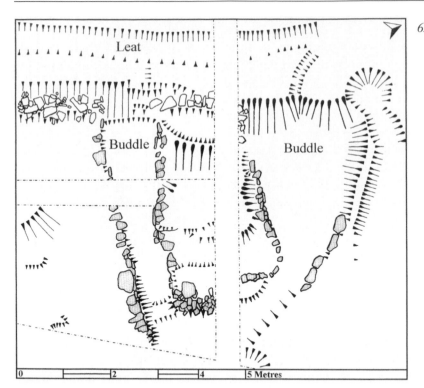

*63 Plan of two buddles excavated at West Colliford. The buddle on the right is very similar in character to one depicted by Agricola (see **62**)*

64 View from the east of the excavations at West Colliford tin mill. In the foreground, a buddle cutting through earlier slime pond deposits can be seen. There was no surface indications of the buddle or slime pond prior to excavation. (David Austin: copyright reserved)

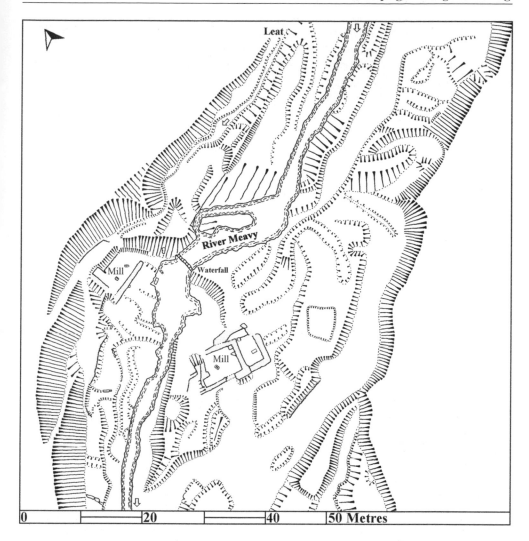

65 Plan of the stamping mills and adjacent streamwork earthworks at Black Tor Falls. Both mills are clearly more recent than the streamwork and would have been used to crush tin ore from the nearby openworks, lode-back pits and shaft mines. The earthworks south west of the western mill probably represent the site of a dressing floor

the feature. The history of the immediate vicinity of a mill is thus one of waste accumulation interrupted by periods of removal where even the most robust structures are likely to have been either buried or destroyed. The excavation of buddles must be considered as an important step towards understanding the efficiency of early dressing techniques. Samples of material collected from strategic positions within these structures should be able to throw light onto this matter. In this respect excavation of the West Colliford buddles was disappointing, since the material analysed was found to contain no tin. However, the potential for this work remains, since at Lanivet a similar investigation of wastes from the twentieth-century mill revealed much useful information on the level of tin recovery, at many stages of the operation. Buddles have been recognised at a small number of sites and amongst these are: Retallack where two possible examples have been identified (**58**); Upper Merrivale where they have been excavated; Black Tor Falls Right Bank (**65**); Lower Yealm Steps (**72**), Ivy Tor Water, Langcombe and Gobbett. As fieldwork continues, no doubt more will be recognised.

8 Smelting

At some stage during the medieval period, the primitive bowl furnaces were replaced by the more substantial blowing houses in which a more permanent stone built furnace was served by bellows operated by a water wheel (**66**). These structures are generally known as blowing houses in Cornwall and blowing mills on Dartmoor. For simplicities sake the term blowing house has been adopted for use in this book because it appears more frequently in the contemporary documentation. The earliest documented example is at Lostwithiel, where in 1332 a 'Blouynghous' was already serving Blackmore stannary. It is, however, likely that these structures were being employed for some time before this date, and they certainly continued in use until the last one in St Austell was abandoned in the 1860s.

Black tin from the streamworks together with that from dressing floors was taken to a blowing house to be smelted. At the blowing house, the tin may have been washed a final time before being put into the furnace together with charcoal. Before the tin could be successfully smelted the furnace had to reach a temperature of at least 1150°c and this was achieved by blowing air through the furnace using water powered bellows. Once the tin had become molten, it flowed from the furnace into a float stone and from here was ladled into a bevelled rectangular trough cut into a granite boulder called a mould stone, in which it cooled to form an ingot of white tin (**Colour Plate 26**).

The blowing house

The most complete contemporary description of tin blowing was given by Thomas Beare. His account is divided into two parts, the first concerning the character of the furnace and the second, the blowing technique. It is too long to repeat here but a number of significant points are raised and certain aspects of it are worth mentioning. Water-powered bellows provided the forced draught necessary for the furnace to operate. Blowing houses were associated with buddles where the black tin was given a final wash and divided into grades. Stream tin was smelted separately from lode tin and, to prevent the latter being driven away by the blast, it was dampened down. The front of the furnace was probably made of clay that was replaced after each use. The furnace was thoroughly heated before the tin was introduced. The position of the bellows nozzle, relative to the furnace was important to ensure a consistent temperature. The order in which black tin was placed into the furnace was coarse stream tin, then fine stream tin, followed by fine load tin and finally waste (slag) from earlier smelting. The tin flowed from the hearth into a trough called a flote. The smelting operation took 15 hours and at least 300lbs (136kg) of white tin were

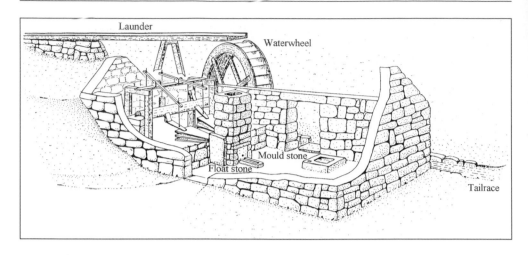

Launder

Waterwheel

Mould stone

Float stone

Tailrace

66 Reconstruction of a blowing house. This illustration highlights the possible layout of a typical blowing house complete with bellows, furnace and mould stone. (Newman, 1998, 37: copyright reserved)

produced every 12 hours in an efficiently managed blowing house. Finally as one might expect the skill of the buddle operator and blower were paramount in achieving the maximum return.

Other writers' accounts although less comprehensive do complement and augment Beare's and thus, for example, we learn from Pryce that the 'flote' which he calls a float stone is a 'moorstone through six feet and a half high, and one foot wide' from which the molten tin was 'laded into lesser troughs or moulds, each of which contains about three hundred of metal'. Richard Carew's account informs us that the tin ingots weighed between 300lb and 400lb (136kg and 181kg) and that each was identified with the owner's mark. The marking of ingots was an obvious deterrent to smuggling and theft, and although only three of the owners' marks are known from this period, those of the blowing houses are more common (**67**). To discourage smuggling and for the purpose of taxation the blowing houses were required to put their mark on the ingots that they produced.

Carew makes it clear that the responsibility for providing the fuel fell on the tinners rather than the blowers, and although he only mentioned wood, peat-charcoal was also employed. Hatcher has been able to demonstrate a close correlation between increased peat exploitation and the tin industry in fourteenth century Helston-in-Kerrier, and by 1446 a shortage necessitated Cornish tinners travelling to Dartmoor for the fuel, a situation which continued into the sixteenth century with Blackmore tinners obtaining at least some peat-charcoal from Dartmouth Forest. Wood charcoal was, however, also extensively used and in 1545 John Pencoste bought part of the coppice wood at Merthen, possibly for use at the Retallack blowing house. Specific charcoal production centres are documented at Hellion in 1605 and 1685 and at Horstocks Wood (Cardinham) in 1701. Whilst in the seventeenth century, Whetter has demonstrated the widespread utilisation of

67 Each blowing house marked the ingots it produced with its own unique mark. Owners of the ingot also marked their ingots. Such measures were taken to prevent theft and other illegal activities. The blowing houses themselves may also have had their mark displayed prominently on the building. (Source: Royal Institution of Cornwall EN. 147)

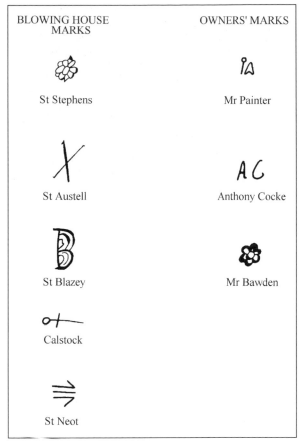

BLOWING HOUSE MARKS	OWNERS' MARKS
St Stephens	Mr Painter
St Austell	Anthony Cocke
St Blazey	Mr Bawden
Calstock	
St Neot	

East Cornwall's woodland for this purpose and Hamilton Jenkin emphasised the importance of charcoal from the New Forest. Both these writers emphasised the importance of importing fuel, but D.B. Barton believed that this was not the case, writing 'that the reputed scarcity of this fuel has hitherto been exaggerated'. To justify this claim he has suggested that scrub woodland, very obvious in the modern landscape, was the most suitable source and that this remained abundant throughout the medieval period. More assuredly he noted that the 'New Blowing House at St Austell, in the period from 1785 to 1807 obtained its supplies of charcoal from woods in a variety of places' within a 10-mile radius. However, by the late eighteenth century most tin was smelted in reverbatory furnaces using coke, and the woodland would have had time to recover from the earlier seventeenth-century excesses.

The quantities of charcoal required to smelt a given amount of tin probably varied according to the efficiency of the furnace, blower and fuel, but Barton believes that a conversion figure of two tons of charcoal to produce one ton of white tin was probably typical. If this figure is accepted, in the first half of the sixteenth century the observed tin output would have required about 1071 tons of charcoal annually to serve the industry. What percentage of this came from peat and wood respectively is not known, but from whatever source, there must have been a sizable charcoal production industry existing simply to cater for the blowing houses. A close correlation between the two industries is most clearly documented by a sixteenth-century writer who stated that 'no man can make his Tynn white without the use of Coale, and therefore are they to be numbered amongst the number of Tynners, for the occupation of the one cannot well stand without the other'

According to the anonymous writer of 1670, alluvial tin was most efficiently smelted with peat charcoal, whilst reworked slag used only wood charcoal and lode tin was smelted with a mixture of the two. Thus in the early medieval period when alluvial tinworks dominated the industry, peat was probably the major source of fuel, but as more lodes were worked, wood became increasingly important. It was thus certainly fortuitous that the major alluvial deposits occurred in association with peat. The rectangular platforms, surrounded by a ditch, on which cut peat was left to dry are abundant throughout the peat production areas of Foweymore, and it is likely that at least some of these are contemporary with the heyday of the blowing houses.

Carew noted that the late sixteenth-century blowing houses only functioned for two or three months, and taking into account time for preparing the furnace and unforeseen obstacles, Greeves has suggested that they may have only been used for about forty days per year. The most plausible explanation for this situation is that white tin was only produced biannually for the midsummer and Michaelmas coinages. Thus if each blowing house was able to produce 300lbs (136kg) per day, 119 such houses would have been needed in Cornwall to produce the known output of 1,390,036 lbs (630,520 kg) in 1521, whilst in Devon Greeves has suggested that 48 could have been responsible for that county's output. At present, only 15 sixteenth-century Cornish blowing houses are known, and the primary value of this admittedly approximate calculation is to demonstrate that many more probably remain to be found by fieldwork and documentary search.

Carew's comments on the working conditions of the blowers are dramatic, but serve to emphasise the harsh nature of the work.

> the blowers' two or three months extreme and increasing labour, sweltering heat, danger of scalding their bodies, burning the houses, casting away the work, and lastly their ugly countenances tanned with smoke and besmeared with sweat (Carew, 1602, 95)

The tinners themselves were not so charitable and generally considered them to be mean. This dislike probably stemmed from the blowers powerful position in the industry. All black tin had to be smelted, and the blowers were in a position to cheat, either by charging extortionate fees, secretly withholding some of the metal or making fraudulent assays of the raw black tin. In addition, blowing was a very skilled operation, little understood by the ordinary tinner, who would have felt aggrieved if the output did not match his expectations, even if it was not the fault of the blower. The blowers were amongst the elite of the working tinners and this difference in social scale could not have helped the general disregard felt by the others. Some blowers even owned their own houses; thus in 1644 John Dale gen. and Walter Hodge are described as owners and blowers 'att the Blowinge howse of St Nyott', but more commonly ownership was vested in big tin adventurers and merchants. The documentation relating to blowing house ownership confirms this contention, and examples include: the influential Enys family who owned the Godolphin Howse in 1662; Penzance merchant John Tremenheere who bequeathed 'the blowing house called Chyandour' in 1686 and John Basset, owner of Tehidy Manor, who in 1656, possessed the Tolcarne blowing house.

In the tin production process the biggest profits could be made in blowing since each house effectively possessed a monopoly of the tin from the surrounding works and the risks were small compared to those searching for and recovering the ore. This situation was confirmed by Carew who claimed that he could not understand how tinners earned enough to survive. The blowers and their employees, however, could not lose; they were provided with the majority of the raw materials by the tinners, and took their fee. The size of the payment to the blowers probably varied according to what the blower thought he might extract from the tinners. The poorer tinners, with little influence, were probably charged more than those of some importance. This failure to set a standard fee probably explains the shortage of details, but an exception is the house at Tolcarne where in 1652 £1.2.6 was charged 'for blowinge one tide and 3 howers' and a further 7s.3d. for 'Meate and drinke', presumably for the blowers.

More enlightening, however, are the details of the costs for blowing at the Lenobray house in 1692. The source of this information is the Enys business ledger of tin loans for the period 1691-8. Eighteen transactions were recorded and in fifteen of these 2s 2d per hour was charged. In the other three instances 1s 6d , 1s 11d and 2s per hour were charged. The standard rate was thus 2s 2d per hour but for some indiscernible reason discounts were occasionally given. In 1692 the Lenobray house operated for at least 1,871 hours (or 78 days). However, this ledger is a very large volume and details of blowing are indicated only under respective accounts. The size of the document and its poor state of repair (particularly towards the end) means that the total number of blowing hours must be considered a minimum figure. Allowing time for maintenance (e.g. 4s. was spent on repairing the furnace, £7 12s 6d for new bellows and 3s for 'oyle for ye bellowes'), clearing of the furnace and unforeseen circumstances, it is obvious that this particular house functioned for more than the two or three months indicated by Carew. It is thus likely that during the seventeenth century at least some of the more popular houses were at work for a longer period. However, blowing houses in less productive areas may have functioned far less frequently during the year.

According to Carew some blowing houses were thatched and tin carried by the force of the bellows into the roof was recovered by periodic burning. By Carew's time this practice was not universal and some furnaces possessed inclined chimneys in which the flying tin metal lodged and could be easily recovered. Loss of tin from the furnace does not appear to have been limited to updraught, and in at least one instance considerable quantities were lost through the base, and into the ground below. In a blowing house at Vellin Antron, near Penryn, about half a dozen pieces of tin metal were believed by the finder to have escaped through a hole in the hearth. Tinners being aware of this problem may have destroyed their furnaces in the search for this metal before finally abandoning a site. This would certainly explain the apparent shortage of surviving furnaces throughout the South-West, despite the relative abundance of the buildings themselves. The anonymous writer of 1670 added very little to Beare's and Carew's accounts, but did note that the furnace employed was called an Alman Furnace and that a special variety of clay was used instead of lime that could not survive the heat. Kaolinised clay would have been able to withstand such temperatures, and therefore, there may have been a small scale industry extracting this material from the granite uplands. The 'claye pyttes' at Steanchegwyn were noted in a bounding document of 1548, and may have been used to provide the furnace linings. Some blowing houses probably prepared their own furnace clay and at Treyew, amongst the many buildings associated with the house was 'one mill for grinding clay + culme'.

On Dartmoor, at least fourteen of the blowing houses also possessed a stamping facility. This probably enabled the blowers to crush and reprocess their slag and at other times to buy undressed ore direct from streamers who did not possess their own stamps because they rarely recovered enough ore to justify possession of crushing facilities. Documentary evidence for mills at which both blowing and stamping were carried out are numerous and include: Bradford Pool; Dartmeet and Lenobray, where the blowing house account for 1692 indicates that five lifters were bought. These accounts also reveal the purchase of stones, and it is possible that these may be tin-stones, which were later crushed by the stamps. Finally at Treyew blowing house, where an inventory notes the presence of 'one double stamping mill and six lifters in good repair' and at Allen, where a blowing house and stamps are recorded in 1690.

At least 148 blowing houses are known and their distribution is shown in figure (**68**). Compared to that for stamping mills (**55**), the distribution is more uniform and there are fewer clearly defined clusters. This is because both alluvial and lode tin were smelted, whereas the stamping mills were largely limited to the immediate vicinity of lodes. Thus on Foweymore, where there are only seven stamping mill sites, there are sixteen blowing houses. Many of these houses must have relied almost exclusively on alluvial cassiterite. Second, some blowing houses are situated in areas where there is no local cassiterite. There are for example, five houses in the immediate vicinity of Truro. The siting of these particular houses is effected by the proximity to the coinage town. Black tin was less cumbersome to handle and transport than white tin and it made sense, therefore, to locate some blowing houses near to the place of coinage, as this would have considerably reduced the distance over which the tin ingots had to be carried before sale. Third, it must be emphasised that it is very unlikely that all the houses represented were contemporary, and particularly where examples exist in proximity it is possible that one was built to replace another. Finally, more specific observations regarding the distribution can be made. The absence of any blowing houses in the western part of Penwith would have necessitated the carriage of all the black tin to the houses situated near the coinage town of Penzance. On Blackmore the majority of the houses are on the southern edges of the moorland. This location probably reflects the need for an abundant water supply and the advantage of proximity to the coinage town at Lostwithiel. The rivers on the moor itself were probably relatively small, whereas those towards the edge were larger and more reliable. This factor probably also explains the Foweymore distribution, where there are no houses in the central part of the moor and the majority are situated in the southern part of the stannary, for easy access to the coinage towns of Bodmin and Liskeard. On

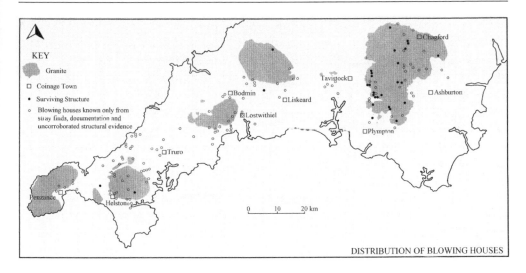

KEY

- Granite
- □ Coinage Town
- • Surviving Structure
- ○ Blowing houses known only from stray finds, documentation and uncorroborated structural evidence

□Chagford

Tavistock□

□Bodmin

□Liskeard

□Ashburton

Lostwithiel

□Plympton

□Truro

Penzance

Helston

0   10   20 km

DISTRIBUTION OF BLOWING HOUSES

68 Map showing the distribution of blowing houses. Most examples survive on Dartmoor, although originally there were once many more in neighbouring Cornwall

Dartmoor the distribution of blowing houses is towards the edge of the granite. The reasons for this probably also relate to the need for a reliable water source combined with ready access to the coinage towns.

Archaeological evidence

This far we have examined the character and nature of smelting tin using some of the available documentation. However, at least 29 blowing houses survive and taken together with the contemporary documentation provide a useful insight into the character of the industry (**69**). Most of the surviving blowing houses lie on Dartmoor and are similar in character to stamping mills but can be differentiated from the former by the presence of a furnace and or by one or more mould stones. They also tend to have a larger ground plan than stamping mills and sometimes fragments of slag may be found.

Furnaces

Furnaces are known to survive at seven sites on Dartmoor (Lower and Upper Merrivale, Upper Yealm, Avon Dam, Gobbett, Taw River and Teignhead). The furnaces are sometimes difficult to determine and survive as small areas defined by large edge set stones (**Colour Plate 27**). The excavation of the furnace at Upper Merrivale revealed a clay lining at the back, but no slag or even evidence that the surviving furnace structure itself had been heated. This suggested that that the internal lining of the furnace had been removed after each smelting and evidence to support this was provided by the discovery of this material elsewhere amongst debris around the building. In addition significant quantities of killas and grey slate in the vicinity of the furnace suggested the presence of some sort of slate superstructure.

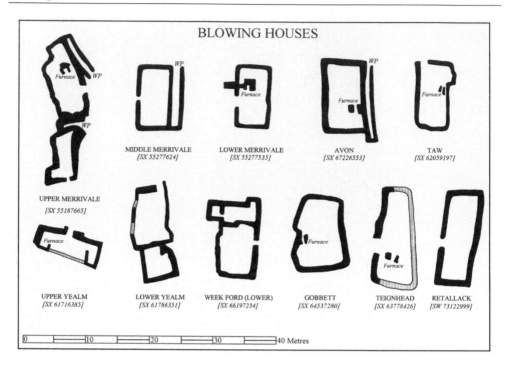

BLOWING HOUSES

MIDDLE MERRIVALE
[SX 55277624]

LOWER MERRIVALE
[SX 55277535]

AVON
[SX 67226553]

TAW
[SX 62059197]

UPPER MERRIVALE
[SX 55187665]

UPPER YEALM
[SX 61716385]

LOWER YEALM
[SX 61786351]

WEEK FORD (LOWER)
[SX 66197234]

GOBBETT
[SX 64537280]

TEIGNHEAD
[SX 63778426]

RETALLACK
[SW 73122999]

0   10   20   30   40 Metres

69 A selection of blowing house plans. Most of these buildings are similar in size, probably reflecting the minimum amount of space required to contain the machinery and equipment

Float stones and mould stones

The tin flowing from the hearth is known to have first accumulated in a float stone from which it was ladled into the mould stone where it cooled to form the ingot. The archaeological evidence relating to this part of the process has been uncovered at a number of sites. No definite examples of float stone have yet been found in Cornwall, but in Devon, examples have been found at Sheepstor Brook, Lower Merrivale and Upper Yealm. The finest example survives at Merrivale where a flat stone with a trough cut into its upper face lies next to the furnace (**Colour Plate 28**).

Mould stones are more abundant with at least 41 examples on Dartmoor and a further three in Cornwall. These stones are generally rough granite blocks with a rectangular trough having bevelled sides and either a flat or concave bottom, cut into the centre (**70**). According to Greeves the most common dimensions are 40 ± 5cm x 30 ± 5cm (15 ± 2 in x 11_ ± 2 in) at the top and 30 ± 5cm x 20 ± 5cm (15 ± 2 in x 7 ± 2 in) at the base, with a depth between 10 and 13cm (4 in and 5 in). There are however, a small number of examples that are either significantly smaller or larger. The larger examples are associated with sites documented as functioning during the later part of the industry's history and would suggest that ingots were more substantial in the later period. This is confirmed by documentation, mainly from coinage rolls, which demonstrates that the Devon ingots increased in size from 100 lbs (45kg) in the late fourteenth century up to 200 lbs (91kg) by 1595. Employing this analogy, the smaller mould stones may have belonged to the

70 *Mould stone at Upper Yealm blowing house. This stone lies just outside the mill where it was discarded*

earlier blowing houses. The shortage of Cornish stones unfortunately meant that it was not possible to conduct a similar exercise, but it is worth noting that ingots varied from 120lb (54kg) in 1305, to between 200 and 250lb (91kg and 113kg) in the late fifteenth century, 345lb (156kg) by the late sixteenth century, from 300 to 400 lbs (136kg to 181kg) in the early seventeenth century and 400 lbs (181kg) by 1641.

The reason for the increasing weight of ingots is unclear, but it does not seem to have been the result of legislation. Certainly, the Stannary authorities probably welcomed the increased weight since the substantial size of the seventeenth-century ingots, in particular, must have discouraged smuggling. One possible explanation is that the ingot weight reflects the amount of tin that could be produced from 12 hours blowing (a tyde). In Beare's account of 1586, the expected outcome from a tyde was 300 lbs (136kg) and ingots produced during this time would certainly have been in keeping with the known average weight for that period. It is thus possible that the differences in ingot weight reflect the increasing efficiency of blowing houses.

This continual increase in ingot size would have required the cutting of new mould stones to accommodate the greater quantities of tin and thus it may be possible to use the moulds to date their associated blowing house. Putting this theory into practise at Retallack, the surviving mould stone, with a volume of 24.12 litres (5.3 gallons) could have produced an ingot with a maximum weight of 388 lbs (176kg) and may therefore have belonged to the early seventeenth century, whilst the two stones from Berriow with volumes of 10.61 litres and 27.7 litres could have produced ingots with weights of 173lb and 445lb (78kg and 202kg) respectively. The first of these ingots may have been used in the early fifteenth century, whilst the second may have been employed in the late seventeenth century.

This approach may also be used to suggest the date of ingots, since any in the range 100 to 450lb (45 to 204kg) must be considered as likely contenders for fourteenth- to seventeenth-century dates. Therefore, smaller ingots were probably earlier, and larger ones of a later date. The smallest ones probably represent the prehistoric examples and this was confirmed by some being found in contemporary contexts, two have been found at the hillforts of Castle Dore and Chun Castle and another at the courtyard village of Porthmeor. The ingots from Carnanton, Trethurgy and Par Beach are the only examples of Roman date, and are some of the few pieces of evidence for working at that time. Seven other ingots weighed between 100 and 150lb (45kg and 68kg) and these probably represent the medieval and early post-medieval examples. Thus, on the basis of weight, the Falmouth ingot at 130lb (59kg) may be seen as fourteenth century, whilst that from neighbouring St Mawes at 158lb (72kg) may be late fourteenth or early fifteenth century, which would certainly support Beagrie's hypothesis that this ingot may be of medieval date. The final group of large ingots consisted of five from Fowey Harbour, varying in weight from 256-368lb (116-167kg). The close proximity of these examples to each other suggested that they were lost at the same time and the variety in weight implies that there was no precisely consistent weight for ingots at any one time, with each blowing house producing its own size of ingots, but within the broad categories of those expected for the period. The Fowey Harbour ingots would thus seem to belong to the early sixteenth century. At least one of these ingots possessed a hole that passed through the block from front to back, and this could have been used to lift the ingot from its mould.

At least seven mould stones with grooves set approximately in the centre of one short side of the mould are known and this may have been used to prevent a stick, which was placed diagonally in the mould, from slipping when the tin was poured in. Once the tin had cooled, the ingot could then be removed using the stick to lift it clear. The evidence from Fowey Harbour indicates that this technique was also employed in Cornwall.

Another feature noted on Dartmoor stones were slight ridges protruding into the base of the mould on the short sides. These would have produced ingots with slight grooves on their under surface and this would have facilitated handling, especially if a rope was used. On some stones a small notch cut into the upper face some distance from the major trough, have been identified as sample moulds. There is, however, no contemporary documentation to support this contention, and Greeves has suggested that they may have been connected with White Rent, a payment in kind, with the tin from the 'sample moulds' being given to the stannary bailiffs. A further possibility is that these moulds were cut to receive illicit tin, since the small size of the resulting ingots would have made their smuggling so much easier. If this was the case, these particular stones must have been in use prior to 1660, when in an attempt to cut down on smuggling the blowing houses became subject to regular supervision by stannary officials, who would not have overlooked such flagrant attempts to produce illegally small ingots.

The production of the tin ingots brings to a close the process we have followed through this book. The story still has some way to go with the ingots being first transported to the nearest coinage town where tax was paid before being bought by merchants who then arranged for their distribution to tin smiths and other craft workers. Tin was widely used in the manufacture of pewter (an alloy of lead and tin) with household utensils, tableware, candlesticks, lamps, caskets, effigies and crucifixes all being made. In addition: tin was an essential constituent of church bells and organs. Tin-rich solders were needed in the building and construction industries, with nails, hinges and iron grills being frequently tinned to prevent rusting, and the metal was also widely used for decoration, with vast numbers of medallions, mementos, brooches and badges being composed entirely of tin. Illuminated manuscripts often contained tin gilding and it was also used to cover wood and other materials as well as being an important constituent of the type-metal used in the printing industry. Bronze continued to be produced and the canon manufacturing and building industries were particularly important consumers of this metal. Tin was thus needed by a variety of developing industries and this must have encouraged continued extraction of the metal, despite ever-increasing exhaustion of the richest and most easily obtainable deposits.

Conclusion

As with many aspects of archaeological and historical research, the detailed observation of the tin industry is simultaneously both absorbing and frustrating. So many elements of the evidence from the earliest to the latest periods are biased or contorted by inconsistencies in the survival of the evidence — both physical and documentary. Certain documents, or landscapes, whilst providing valuable insights into many aspects can, at the same time, completely bias our understanding of the greater picture. Why is it, that the documentary evidence so often relates to destroyed sites, or rich landscapes tantalisingly, to date at least, remain divorced from their documentary proof? So, we are adrift amidst a sea of evidence from which we may imply, suppose, interpolate and extrapolate, in order to make sense of the greater whole. Will the complete truth ever emerge? will there be a point at which we can categorically state the whole truth of the evidence we see? probably not. However, future research, exploration, chance discovery and undoubtedly more frustration will eventually lead to a fuller and more comprehensive understanding of what is a highly complex subject. Whatever an individual sees as being a crucial part of this industry — whether it be legal, social, economic, environmental or purely technological, it still forms a critical part of a complex whole waiting to be fully understood. Let us await with anticipation the future revelations, set our sights to a horizon in which comprehension clears the picture, enabling us to view the whole with clarity and understanding and above all to enjoy all our endeavours in achieving this result.

9 Sites to visit

The archaeological remains associated with the early tin industry survive to some extent within most of the stanniferous parts of the south-west of England. In places, especially in Cornwall later mining activity has often caused severe damage, but despite this, even in these locations traces of the earlier features and structures do survive. They are, however, much more difficult to decipher and therefore, it is much easier to study the evidence in places where it survives best. In general terms, Dartmoor, with its large areas of public access, combined with excellent quality of archaeology provides many of the best preserved and intelligible sites. Most of the sites outlined below benefit from public access, but some are in private hands. Access arrangements can vary over the years and I would, therefore, strongly recommend that local inquiries are made before entering enclosed land. Mining sites of all ages are dangerous and care should be taken at all times. Many tinworks are seasonally obscured by bracken. It is therefore recommended that, in order to avoid disappointment, visits should be carried out between the end of December and the start of June.

This selection is designed to provide an introduction to the types and character of the archaeological evidence surviving in the field. They are amongst the best preserved and together form an insight into the varied and interesting character of the evidence on which much of this book relies (**71**).

Alluvial streamworks
Brim Brook (SX 589874) (14 & Colour Plate 12)
This streamwork contains a range of well-preserved earthworks. Most of the waste dumps lie at right angles to the valley bottom and many of them are revetted by drystone walling on the side facing upstream. Three discrete blocks of earthworks are visible and each is served by a leat. Within the streamwork there are three buildings used by the tinners for shelter and storage.

Erme Valley (SX 639650)
The largest single block of well-preserved alluvial streamworks survives within the Erme valley and extends from NGR SX 641626 to SX 624678. Particularly fine examples of retained dump earthworks survive. Over 30 tinners' buildings and at least five stamping mills are also known.

Langcombe Brook (SX 602670) (26)
A series of well-preserved stone revetted dumps survive within this streamwork. The revetment faces downstream and this is one of the areas where it has been demonstrated

71 Location of sites to visit

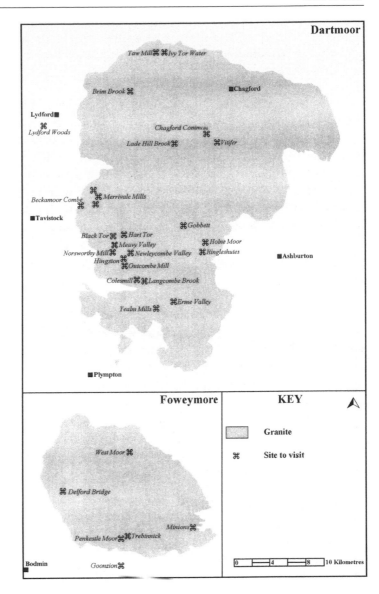

that reworking of tin deposits was sometimes carried out in a downstream direction. Several tinners' buildings lie within this streamwork.

Meavy Valley (SX 574715) (20, 23, 25 & 42)
A large well-preserved streamwork survives along the upper reaches of the River Meavy and Hart Tor Brook. The distinctive linear banks can be easily seen and whilst they may seem somewhat confusing when viewed from ground level this is because the picture has been scrambled by phases of reworking being carried out in several different directions. On the Hart Tor Brook, in particular, it is possible to identify two major phases of activity. The earliest one has left behind substantial rounded banks, which generally lie parallel with the valley bottom. The later phase is represented by narrower dumps, which are

sometimes revetted and generally lie at an angle to the valley bottom. It is tempting to equate the earliest dumps with the medieval period and the others to sometime during the post-medieval times.

Eluvial Streamworks
Beckamoor Combe (SX 535754) *(Colour Plate 10)*
This streamwork lies adjacent to a small car park and is therefore readily accessible. Survey work here by the Dartmoor Tinworking Research Group identified several phases of exploitation. At the upper end of the streamwork is a small tinners' building complete with fireplace. South of the road leading across the streamwork there are two medieval farmsteads. The western one has been partly cut by the streamwork, but the other settlement survives intact. A field boundary leading from this farmstead runs to the edge of the streamwork before turning to run alongside it for a short distance. This relationship implies that the streamwork was already in existence before this medieval wall was constructed.

Ivy Tor Water (SX 628916) *(56)*
The streamwork at Ivy Tor Water (or Ladybrook) can be best appreciated from the slopes of Cosdon Hill. The marked curve of the banks can be easily traced. Descending down to the streamwork the remains of a small stamping mill complete with buddles can be found at SX 62859175. There is no evidence of any mining activity within the vicinity and it may therefore be presumed that large pieces of cassiterite which were encountered during streamworking were processed at this site.

Newleycombe Valley (SX 586704) *(42)*
On the northern side of the Newleycombe Valley there are four streamworks. The streamworks all possess parallel dumps which lie across the contour. The streamwork at NGR SX 59357016 has been partly cut through by a later openwork. Reservoirs and leats which were so essential for the process survive within the vicinity of all the streamworks.

Penkestle Moor (SX 178699) *(12, 13 & 41)*
The streamworks on Penkestle Moor illustrate the hydrological skills of the tinners. A complex system of leats and reservoirs were used to carry and store the water before it was used in the streaming process. Cutting into part of the streamwork are a line of later lode-back pits.

Shoad works
Goonzion (SX 176675) (33)
The surface of Goonzion Downs resembles a lunar landscape with thousands of small pits cutting into the hillside. These were excavated by tinners searching for ore. The sheer quantity and density of pits indicates that they were not used for prospecting, and tin must have been removed and taken for processing elsewhere. This type of deposit would normally have been worked using streaming techniques, but because of the difficulties in

bringing water to this site, shoad collection pits had to be excavated instead. The larger pits within the area are of the lode-back variety and the linear gulleys represent prospecting trenches. Three separate tinworks were documented at Goonzion in 1516.

Kit Hill (SX 375713) (34)

The upper slopes of Kit Hill are covered in shoad collection pits. However, the picture is complicated by the survival of a streamwork, lode-back workings, prospecting pits and trenches together with a great many other features relating to later shaft mining.

Lode-back Tinworks
Black Tor (SX 573716) (9 & 35)

The small lode-back tinwork on Black Tor is readily accessible from the nearby road. It consists of lines of pits dug onto the back of the lode, together with other smaller pits which were used to first find and then examine the lodes in detail. At the lower end of the tinwork there are two adits which may imply that at least some of the pits were drained. The tin from this tinwork was probably stamped in the nearby Black Tor Falls stamping mills.

Carn Brea (SW 686407)

Cutting through the earlier Neolithic hilltop enclosure are at least three lines of lode-back pits. Many of the pits are substantial and may indeed have been proper deep shafts. Two whim platforms surviving within the area confirm that at least some shaft mining was also carried out.

Hingston (SX 583691) (42)

The lode-back tinwork on Hingston Hill forms part of the group in the Newleycombe Valley. It partly cuts through a prehistoric settlement and runs very close to a group of prehistoric cairns and stone row.

Holne Moor (SX 677711)

On Holne Moor (an area more renowned for its very fine prehistoric archaeology) there are several lode-back tinworks. Some of these cut through the earlier prehistoric fields but most lie to the south at Ringleshutes.

Openworks
Hart Tor (SX 577719) (42)

The openworks at Hart Tor can be seen clearly from the main road leading between Princetown and Yelverton. This site is unusual in that a large number of interconnected gulleys were cut to form the whole. A large number of leats surviving within the vicinity of this tinwork are believed to have been connected with prospecting rather than exploitation.

Vitifer (SX 688808) (43)

The area surrounding the Warren House Inn contains more openworks than any other.

Here a large number of lodes have been exploited using opencast quarries.

Most of the shafts surviving in this area probably belong to later periods of mining but some may have had early origins.

Shaft and Adit Mines
Black Tor (SX 57367150) (9)

Two adits situated close to each other lie just above the River Meavy at NGR SX 57367150. These adits would have drained and provided access to the tinwork on the southern slopes of Black Tor. The lode which these adits served had previously been worked using lode-back pits.

Chagford Common (SX 67778270) (50)

The shaft on Chagford Common at NGR SX 67778270 is associated with a whim platform on which winding machinery to raise ore and rock would have operated. In the centre of the platform is a melior stone in which the winding drum rotated. Some spoil from the shaft was dumped into the nearby openwork, indicating that this tinwork was already abandoned before the shaft was dug.

Trebinnick (SX 183704) (14 & 49)

This small tinwork includes two shafts, an adit, at least three shelters and a small number of prospecting pits. The absence of a stamping mill probably means that it represents the site of a prospecting venture.

Tinners' Buildings
Delford Bridge (SX 11487575) (14)

This small earthwork structure is probably an example of a tinners' shelter built entirely from turf

Lade Hill Brook (SX 63928145) (Colour Plate 4)

This small circular structure with corbelled walling is, because of its shape, known as a beehive hut. This building lies within an alluvial streamwork and may have been used as a shelter for a small number of tinners or more probably mainly for storage.

Stamping mills
Black Tor Falls (SX 574716) (65, Front Cover & Colour Plates 17 & 29)

The two stamping mills at Black Tor Falls sit below a picturesque waterfall and are a popular destination for visitors. The mill on the eastern bank is best preserved with the door lintel still in position. At least two mortar stones lie within this mill, whilst five lie in or close to the one on the opposite side of the river. Earthworks adjacent to the western mill may be the remains of buddles.

Colesmills (SX 59376676) (56)

The wheelpit is the most obvious feature at this mill. A mortar stone has been reused to carry the axle of the waterwheel and this still sits on one edge of the wheelpit. The leat serving this mill can be traced leading northward and earthworks in the vicinity of the building may represent the remnants of a dressing floor.

Norsworthy Left Bank (SX 56786958) (56 & Colour Plates 16 & 20)
The mill on the eastern bank of the River Meavy near Norsworthy nestles at the foot of a steep slope. Within and around the building there are six mortar stones and two slotted stones which may have been used to support the machinery.

Outcombe (SX 58016860) (56)
Lying within and around this mill are at least 21 mortar stones. Apart from the mill building itself, which stands up to 1.9m high, there is an obvious wheelpit and tailrace as well as leat and embankment. A small structure lying immediately to the north of the mill may be a buddle.

Crazing mills
Gobbett (SX 64537280) (69 & Colour Plate 23)
The mill at Gobbett is unusual because mortar stones, grazing stones and mould stones all survive within the building. This important site therefore contains archaeological information relating to all tin processing methods.

Retallack (SW 73182979) (58 & Colour Plates 18, 24 & 26)
The finest collection of tin mills of all three types survives at Retallack, where at least two stamping, four crazing and one blowing house lie together within a beautiful wooded valley. Eleven mortar stones confirm that stamping once played a part in this very important complex. The northernmost mill was also associated with crazing mill stones. However, the main concentration of mortar stones was associated with the southern part of the complex. Here the distribution of stones was largely limited to the upper building and the hillside immediately downhill of the northern doorway. The picture that emerges is of stones being thrown out of the entrance and rolling downwards towards the stream.

Blowing Houses
Merrivale Group (69 & Colour Plates 21, 27 & 28)
The finest collection of surviving blowing houses lies adjacent to the River Walkham north of Merrivale. The three mills have very similar ground plan dimensions and each contains at least one mould stone. The lower mill (SX 55277535) contains a well-preserved furnace complete with float stone. The middle mill (SX 55277624) lies adjacent to a medieval longhouse and field system. The upper mill (SX 55187665) which has been excavated by the Dartmoor Tinworking Research Group was converted from a stamping mill into a blowing house sometime in the early part of the eighteenth century. Large numbers of mortar stones found during the excavation indicate that this site was also an important stamping and processing area.

Taw (SX 62059197) (69)
This large building nestles at the foot of a steep scarp next to the River Taw. Two set slabs

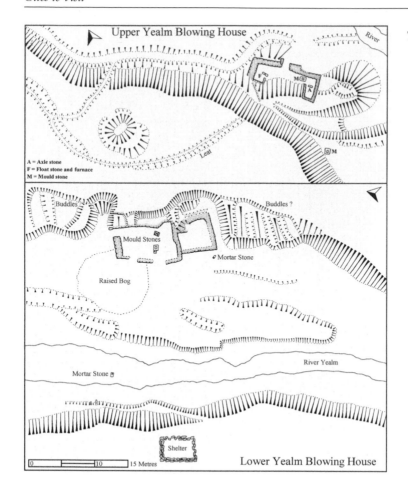

72 Plan of the two blowing houses in the Yealm Valley. Both buildings contain mould stones and the lower one also possessed stamping machinery as witnessed by the mortar stones and buddles. The Upper mill has a furnace and the eastern side of the building was converted into a small tinners' shelter

within the building may represent a furnace, although more conclusive evidence to support its blowing house identification comes from the large quantities of slag that have been recovered from the site. A very fine walled tail race leads away from the building

Yealm *(70 & 72)*

The two blowing houses on the Yealm survive within earlier streamwork earthworks. The Lower Yealm blowing house (SX 61786351) was investigated by Kelly in 1866 and unusually contains mould stones each containing two troughs. Two mortar stones and earthworks which may be buddles indicate that stamping was also carried out here. The Upper blowing house (SX 61716385) contains a furnace, mould and axle stones. The eastern part of this building was converted into a tinners' shelter in later years.

Landscapes

Minions *(SX 2571) (73)*

Lines of lode-back pits dominate the landscape around Minions, but streamworks and shafts also attest to the considerable mineral wealth of this area. Later engine houses and dressing floors indicate that extraction continued into the modern period.

73 Tinworks near
Minions. The deep
gash in the centre of
this photograph is an
eluvial streamwork
measuring up to 10m
deep. Elsewhere, lines
of lode-back pits can
be clearly discerned.
Some of these pits cut
into the base of the
earlier streamwork. In
the foreground a shaft
with associated whim
platform cuts through
earlier lode-back pits.
(Photograph by Steve
Hartgroves, Cornwall
Archaeological Unit:
copyright reserved)

Newlcycombe Valley (SX 5970) (42)
A comprehensive range of archaeological remains connected with the tin industry survive within this valley. Good examples of most types of site are to be found and a full day spent in the field here will provide an excellent introduction to the character and nature of the industry.

Ringleshutes (SX 674698)
An impressive array of streamworks, openworks and lode-back tinworks lie within this area. Whilst some of these features are undoubtedly modern, together they give the impression of a landscape which has been extensively worked using the different extraction techniques .

West Moor (SX 1880) (74)
A large number of splendid streamworks can be seen on this part of Foweymoor. In places

74 Streamworks on West Moor. In the centre of the photograph is an alluvial streamwork complete with parallel dumps. The gash cutting into the hillside in the distance may look like an openwork, but the presence of distinctive parallel spoil dumps in its base indicates that it is an eluvial streamwork. A small adit and dump are visible at the foot of the photograph. (Photograph by Steve Hartgroves, Cornwall Archaeological Unit: copyright reserved)

the streamworks are up to 10m deep indicating the huge amounts of material that have been removed. Many of these tinworks are shown on the 1:25000 Ordnance Survey Map of the area.

Glossary

Adit A level tunnel driven into the hillside to facilitate access, drainage and haulage of ore to the surface. The term 'drift' is sometimes used to describe an adit.

Alluvial streamwork Streamworks found lying beside rivers are known as alluvial streamworks because they were exploiting tin deposits previously deposited by the river.

Blowing house Building in which tin was smelted with charcoal in a stone built furnace served by bellows operated by a water wheel. These buildings are also known as blowing mills.

Buddle Rectangular or triangular sloping box-shaped structure in which cassiterite was separated from lighter wastes using water.

Cache A small stone-built structure in which tools and black tin were hidden. Many caches were built into earlier waste dumps.

Cassiterite Tin oxide, also known as black tin.

Crazing mill Building in which crushed ore was reduced to a consistency where it could be efficiently dressed. The machinery consisted of two horizontally placed millstones sitting within a frame. The ore was fed in through a hole in the top stone, into the space between the two, and then ground to a fine powder by the upper stone rotating against the lower stationary one.

Eluvial streamwork Eluvial streamworks exploited deposits of cassiterite which had been detached from the lode and exposed to weathering and often transportation, but had not been sorted by alluvial action. They most commonly occur in dry shallow valleys lying above the larger rivers, but in some instances because of the character of the local topography, the eluvial deposits may lie immediately above their parent lode.

Float stone The stone into which molten tin poured from the furnace. Surviving examples include a shallow trough cut into a granite slab.

Gangue Minerals other than cassiterite found in tin ore.

Hatch A pit.

Ingot A block of white tin.

Leat An artificial channel to carry water.

Level A drainage channel carrying waste in suspension from a streamwork.

Lode A linear area of mineralisation within the underground rock.

Lode-back pit A pit dug onto the upper part of a lode, from which tin ore was extracted.

Melior stone Stone sitting in the centre of a whim platform in which the vertical axle of winding machinery rotated. Melior stones all have a circular hole cut into their upper face and this shows traces of wear caused by the rotating action of the winding drum.

Mine Generic term given to any tinwork extracting tin ore directly from the lode. Types of mine include: shaft and adit, openwork and lode-back pit tinwork.

Mortar stone A stone on which ore was crushed by mechanically driven stamps. The continual pounding of the stamps resulted in the formation of distinctive small saucer shaped hollows.

Mould stone A rough granite block with a rectangular trough having bevelled sides and either a flat or concave bottom, cut into the centre. Molten tin from the float stone

Openwork Also known as beams (mainly on Dartmoor) and coffins (Cornwall). They were formed by opencast quarrying along the length of the lode. Openworks generally survive as deep, narrow and elongated gulleys.

Prospecting pit Pit excavated during the search for tin. The pits themselves survive individually as small rectangular or oval hollows with an associated crescent-shaped bank, normally lying downslope of the pit.

Shaft A deep, vertical or near-vertical tunnel leading from the surface to provide access to underground workings.

Shoad Tin ore detached from its lode.

Stamping Mill Building in which stamping machinery was employed to crush tin ore. The name derives from the use of water-driven stamps to pound the ore.

Stannary A district under the jurisdiction of a legal and administrative organisation, the Stannary Court, which passed laws, protected the privileges of tinners and supervised the collection of taxes on tin presented for coinage. The stannary towns in Cornwall were Bodmin, Helston, Liskeard, Lostwithiel, Penzance and Truro whilst in Devon they were Ashburton, Chagford, Tavistock and Plympton.

Tin Mill A building in which machinery powered by one or more waterwheels was used to process tin ore. Stamping and crazing mills crushed the ore and blowing houses smelted it.

Tinners' Building Generic term used to describe all buildings erected by the tinners, other than mills.

Tinbound A legal claim to extract tin from a specified area.

Tinwork A term which covers all the different types of tin extraction including streamworking and mining.

Stope This term has two different uses. In an openwork, it refers to steps or platforms that were formed in the base of the quarry during extraction. Underground, the term refers to the void left behind after the lode has been removed.

Streamwork Tinwork in which water was used to separate the cassiterite from previously weathered deposits.

Tye This term has more than one meaning. In streamworking, it is the name given to the work area where the first stage of hydraulic separation is carried out. On dressing floors it is the name of a piece of equipment used to dress the tin. Whilst at some mines the term tye was used to describe an adit or drainage level.

Wheelpit Rectangular stone-lined structure in which a wheel rotated. Wheelpits are most commonly found associated with stamping mills and blowing houses.

Whim platform An artificially created circular platform on which a winding drum powered by horses was situated. Whim platforms are invariably situated adjacent to a shaft.

Further reading

A Manuscript Sources

A substantial number of documents were consulted during the preparation of this work. A full list appears in Gerrard, (1986) and many of the particularly significant ones appear within the text of this book. In particular, I would wish to draw attention to the extremely useful Calendars produced by Charles Henderson and now held by the Royal Institution of Cornwall in Truro.

B. Printed Sources

Agricola, G., *De Re Metallica* (ed. H.C. & L.H. Hoover),1950 edn. (New York) Originally published in 1556.

Alcock, L., , *Arthur's Britain — History and Archaeology A.D. 367-634*, (Penguin) 1971.

Anon., 'An Accompt of some Mineral Observations touching the Mines of Cornwall and Devon; etc.', *Philosophical Transactions of the Royal Society of London*, no. 5, 1670, pp. 2096-2113.

Austin, D., Gerrard, G.A.M. & Greeves, T.A.P. 'Tin and agriculture in the middle ages and beyond: landscape archaeology in St Neot Parish, Cornwall', *Cornish Archaeology*, no. 28, 1989, pp. 5-251.

Barton, D.B., *Essays in Cornish Mining History*, 2, Truro, 1971.

Beagrie, N., 'Some early tin ingots, ores and slags from Western Europe', *Historical Metallurgy,* no. 19, 1985, pp. 162-8.

Blanchard, I., 'The Miner and the Agricultural Community in Late Medieval England', *Agricultural History Review*, no. 20, 1972, pp. 93-106.

Blanchard, I., 'Rejoinder: Stannator Fabulosus', *Agricultural History Review*, no. 22, 1974, pp. 62-74.

Borlase,W., *The Natural History of Cornwall* (0xford) 1758.

Brown, A.P., 'Late Devensian and Flandrian Vegetational History of Bodmin Moor', *Philosophical Transactions of the Royal Society of London B.*, no. 276, 1974, pp. 251-320.

Bryant, J., 'On the remains of an ancient crazing mill in the parish of Constantine', *Journal of the Royal Institution of Cornwall*, no. 7, 1882, pp 13-4.

Burnard, R., 'On the Track of the 'Old Men' of Dartmoor. Parts 1 & 2', *Transactions of the Plymouth Institute*, no. 10, 1887-90, pp. 95-112 and 223-42.

Butler, J., *Dartmoor Atlas of Antiquities*, (4 Vols), Devon Books, 1991-1994.

Carew, R., *Survey of Cornwall*, 1969 edn (ed. F.E. Halliday). Originally published in 1602.

Collins, J.H., *Principles of Metal Mining*, 1875.

Costello, L.M., 'The Bradford Pool Case', *Rep. Trans. Devonshire Ass.*, no. 113, 1981, pp. 59-77.

Earl, B., *Cornish Mining*, Truro, 1968.

Fiennes, C., *The Illustrated Journeys of Celia Fiennes c.1682–c.1712* (ed. C.Morris), (London), 1982 edn.

Finberg, H.P.R., *Tavistock Abbey — A Study in the Social and Economic History of Devon*, Cambridge University Press, 1951.

Fleming, A., 'The Dartmoor reaves: boundary patterns and behaviour patterns in the second millennium BC', *Devon Archaeological Society Proceedings*, no. 37, 1979, pp. 115-131.

Fleming, A., 'The prehistoric landscape of Dartmoor: wider implications', *Landscape History*, no. 6, 1984, pp. 5-19.

Fox, A., 'Excavations on Dean Moor, in the Avon Valley, 1954-1956', *Rep. Trans. Devonshire Ass.*, no. 89, 1957, pp. 18-77.

Fox, A., 'Tin Ingots from Bigbury Bay', *Devon Archaeological Society Proceedings*, no.53, 1995, pp. 11-23.

Fox, H.S.A., 'Medieval Dartmoor as Seen Through its Account Rolls', *Devon Archaeological Society Proceedings*, no. 52, 1994, pp.11-23.

Gerrard, G.A.M., 'Retallack: a late medieval tin mining complex in the parish of Constantine, and its Cornish Context', *Cornish Archaeology*, no. 24, 1985, pp. 175-82.

Gerrard, G.A.M., *The early Cornish tin industry an archaeological and historical survey*, Unpublished Ph.D. Thesis, University of Wales, 1986.

Gerrard, S., 'Streamworking in medieval Cornwall', *Journal of the Trevithic Soc.*, no.14, 1987, pp. 7- 31.

Gerrard, S., 'The Beckamoor Combe Streamwork Survey', *Dartmoor Tinworking Research Group Newsletter No. 3*,1992, pp. 6-8.

Gerrard, S., *Meavy Valley Archaeology, Site Report No. 10, Hart Tor Tinworks*, 1998.

Godley, A.D., *Historiae* (4 volumes),1960.

Greeves, T.A.P., *The Devon Tin Industry 1450 — 1750 An archaeological and historical survey*, Unpublished Ph.D. Thesis, University of Exeter,1981a.

Greeves, T.A.P., 'The Archaeological Potential of the Devon Tin Industry', in Medieval Industry (ed D.W. Crossley), *CBA Research Report 40*, 1981b, pp. 85-95.

Greeves, T.A.P., 'The Dartmoor Tin Industry — Some aspects of its field remains', *Devon Archaeology*, no. 3, 1985, pp. 31-40.

Hamilton Jenkin, A.K., *The Cornish Miner — An account of his life above and underground from Early Times*, 1927.

Hatcher, J., *Rural Economy and Society in the Duchy of Cornwall, 1300-1500*, Cambridge, 1970.

Hatcher, J., *English Tin Production and Trade before 1550*, Oxford,1973.

Hatcher, J., 'Myths, Miners and Agricultural Communities', *Agricultural History Review*, no. 22, 1974, pp. 54-61.

Henwood, W.J., 'On the detrital tin-ore of Cornwall', *Journal of the Royal Institution of Cornwall*, no. 4, 1874, pp. 191-254.

Herring, P., *Godolphin, Breage — An Archaeological and Historical Assessment*, Cornwall Archaeological Unit, 1998.

Hitchens, F. & Drew,S., *The History of Cornwall from the Earliest Records and Traditions, to the Present Time*, 2 vols., 1824.

Hunt, R., *British Mining — A Treatise on the History, Discovery, Practical Development and Future Prospects of Metalliferous Mines in the United Kingdom*, London, 1887.

Kelly, J., 'Celtic Remains on Dartmoor', *Rep. Trans. Devonshire Ass.*, no. 1, 1866, pp. 45-8.

Leifchild,J.R., *Cornwall its Mines and Miners*, 1855. Reprinted 1968, New York.

Lewis, G.R., *The Stannaries — A Study of the Medieval Tin Miners of Cornwall and Devon*, 1908. 1964 edn.Truro.

Maclean, J., 'The tin trade of Cornwall in the reigns of Elizabeth and James compared with that of Edward I', *Journal of the Royal Institution of Cornwall*, no. 15, 1874, pp. 187-190.

Merret, C., 'A Relation of the Tinn-Mines, and working of Tinn in the County of Cornwall', *Phil. Trans. Roy. Soc. London*, no. 138, 1678, 949-952.

Muhly, J.D., *Copper and Tin: The distribution of mineral resources and the nature of the metals trade in the Bronze Age*, Ph.D. Thesis, Yale University, 1969.

Newman, P., 'The Moorland Meavy — a Tinners' Landscape', *Rep. Trans. Devonshire Ass.*, no. 119, 1987, pp. 223-40.

Newman, P., 'Week Ford Tin Mills, Dartmoor', *Devon Archaeology Society Proceedings*, no. 51, 1993, pp. 185-197.

Newman, P., *The Dartmoor Tin Industry — A Field Guide*, Chercombe Press, 1998.

Norden, J., *Speculi Britanniae Pars; a topographical and historical description of Cornwall* 1584, (1728 edn.).

Northover, J.P., 'The Exploration of the Long-distance Movement of Bronze in Bronze and Early Iron Age Europe', *Bulletin of the Institute of Archaeology*, 1982, pp. 45-72.

Penhallurick,R.D., *Tin in Antiquity*, Institute of Metals, 1986.

Pennington, R.R., *Stannary Law — A History of the Mining Law of Cornwall and Devon*, David & Charles,1973.

Price, D.G., 'Changing perceptions of prehistoric tinning on Dartmoor', *Rep. Trans. Devonshire Ass.* no.117, 1985, pp. 129-38.

Price, D.G., 'Prehistoric tin extraction on Dartmoor: a reply to Andrew Fleming' *Rep. Trans. Devonshire Ass.* no.120, 1988, pp. 91-5.

Pryce,W., *Mineralogia Cornubiensis*, 1778, (1972 edn.).

Sharpe, A., *The Minions Area — Archaeological Survey & Management*, 2nd edn., Cornwall Archaeological Unit, 1993.

Shell, C.A., 'The early exploitation of tin deposits in south-west England', in *The Origins of Metallurgy in Atlantic Europe,* M.Ryan (ed.), 1978, pp. 251-63.

Simmons, I.G., 'An Ecological History of Dartmoor', In *Dartmoor Essays*, I.G.Simmons (ed.) 1964.

Stocker, H.M., 'Account of some remains found in Pentuan streamwork and of the circumstances under which they were found', *Penzance Natural History and Antiquarian Society*, no. 2, 1852, pp. 88-90.

Threipland, L.M., 'An Excavation at St. Mawgan-In-Pydar, North Cornwall', *Archaeological Journal*, no. 93, 1956, pp. 33-81.

Tylecote, R.F., *Metallurgy in Archaeology*, 1962.

Tylecote, R.F., 'Furnaces, Crucibles and Slags', in *The Coming of the Age of Iron* (ed. T.A.Wertime & J.D. Muhly), 1980.

Wacher, J., *Roman Britain,* 1978.

Webster,J., *Metallographia: An History of Metals*, 1671.

Whetter, J., *Cornwall in the Seventeenth Century — an economic history of Kernow*, 1974.

Worth, R.H., 'The Stannaries', *Transactions Plymouth Institution*, no.15, 1910, pp. 21-45.

Worth, R.H., *Worth's Dartmoor*, 1981 (3rd impression).

Index

Numbers within brackets refer to the National Grid Reference
Numbers in bold refer to the page numbers on which relevant illustrations can be found

Index of Places

Aegean 20
Allen, blowing house and stamping mills (SW 831496) 134
Arrows Flight (SX 1771) 33, **34**
Ashburton **11**, 104, 152
Avon, blowing house (SX 67226553) 135, **136**
Baldew, mine (SX 0050) 95
Baldue, mine (SW 7742) 22
Ball West, mine (SW 5930) 29, 36-7, 42, 95-6
Ballowall, smelting site (SW 356313) 19
Bartinney, lode-back tinwork (SW 394296) 83
Beame, Greate, tinwork (SX 0157) 36
Beckamoor Combe, eluvial streamwork (SX 535754) 76, **141**, 142, **colour plate 10**
Beckamoor Combe, tinners' building (SX 53537569) 142, **colour plate 3**
Belerium 21
Belowda, shoad work (SW 970624) 79
Beme, The, openwork in Roche 92
Berriow, blowing house (SX 272757) 137
Bigbury Bay, ingots from (SX 6046) 20
Birch Tor, openworks (SX 682814) 91
Black Ridge Brook, tinners' building (SX 59358449) **49**
Black Rock, tinwork at (SX 194797) 27, **28**
Black Tor Falls, left bank, stamping mill (SX 57497161) **29**, **86**, **127**, **141**, 144, **front cover, colour plates 17 and 29**
right bank, stamping mill (SX 57487162) **29**, **86**, **127**, 128, **141**, 144
Black Tor, adits (SX 57367150) **86**, 96, **141**, 143, 144
tinwork (SX 573716) 27, 28, **29**, **82**, **86**, 96, **141**, 143, 144
Blacketor Combe, tinbound (SX 1573) 43
Blacklane Brook, alluvial streamwork (SX 628674) 72
Blackmore, stannary 32, 35, 36, 56-8, 61, 62, 129, 130, 134
Bodmin **11**, 134, 152
Bolster Bank (SW 7149) 19
Boscarne, Roman tinwork (SX 039675) 21-22
Boscundle, stamping mill (SX 045531) 44
Bradford leat (SX 657906) 47

Bradford Pool, blowing house and stamping mills (SX 699910) 134
Breedwell in Perranzabuloe, tinbound 35
Brim Brook, alluvial streamwork (SX 589874) 72, 140, **141**, **colour plate 12**
tinners' building (SX 58728690) **49**, 140
tinners' building (SX 59038767) **49**, 140
tinners' building (SX 59098781) **49**, 140
Brittany 14
Broadwater Moor, Luxulyan, Bronze Age vessels from 15
Brode open worke, tinwork (SX 1569) 43
Brodhok, streamwork 61
Brown Gelly, tinners' building (SX 20097280) **49**
whim platform (SX 193724) **99**, 101
Bunning's Park (SX 18377189) 55, **colour plate 7**
Burlazack al's Bolazack, tinwork (SX 0157) 36
Buttern, alluvial streamwork (SX 183812) 70, **148**
Caer Bran, lode-back tinwork (SW 406292) 83
Camerton 22
Carbous Stamps, stamping mill (SW 55303359) 111
Carbusse Worck, tinwork (SX 0157) 36
Carclase Ball, tinwork (SX 0254) 44
Carloggas, enclosure (SW 874654) 19
Carn Brea, hilltop enclosure (SW 686407) 19, 143
Carnanton, ingot from (SW 8865) 22, 138
smelting site (SW 8865) 19
Carne, alluvial streamwork (SX 199826) 70
eluvial streamwork (SX 201811) **74**, 76
Carnon, finds from (SW 7839) 15
Castallack roundago (SW 448254) 19
Castle Dore, ingot from (SX 103547) 138
Castle Gotha (SX 027496) 22
Castle Park, dressing floor at (SX 1060) 124
Chagford **11**, 152
Chagford Common, whim platform (SX 67778270) **99**, **141**, 144
Challacombe, openworks (SX 693805) **colour plate 15**

Charterhouse 23
Cheesewring, mine at (SX 258722) 99
Chun Castle, hillfort (SW 405339) 23, 138
Chun Downs, lode-back tinwork (SW 403335) 83
Chyandour, blowing house (SW 4731) 133
Chysauster, Iron Age village (SW 472350) 19
Clennacombe Streme and beme worke (SX 2772) 38, 62
Clodgy, bank (SW 5140) 19
Clowance, tin house (SW 6334) 49
Codda, eluvial streamwork (SX 174792) 79
Colesmills, stamping mill (59376676) **141**, 144-5
Colliford, East, cache (SX 17837167) **33**, 50, **colour plate 6**
    stamping mill (SX 179717) 120
    reservoir (SX 1771) 12
    streamworks (SX 178716) 23, 32, **33**, 35, 55, 61, 64-7, **117**
    West, openwork (SX 175711) **89**, **90**, 91-2, **117**
    West, stamping mills (SX 17777121) 107, **114**, 116, **117-9**, 120, 123, 125, **126**, 127, **colour plates 22 and 25**
Combe Martin Silver Mines 93
Cosgarne, crazing and stamping mills 120
Coweswerk in St Buryan, streamwork 61
Crugebras, tinwork (SW 7443) 39
Cut Combe Water, tinners' building (SX 59148365) **49**
Dartmeet, blowing house and stamping mills (SX 67277355) 134
David's Church, tinwork (SX 1569) 43
De Quelhiocke in Blackmore, tinwork 39
Deadmans Bottom, tinners' building (SX 60726674) **49**
Dean Moor, slag and pebble from (SX 677654) 19
Delford Bridge, tinners' building (SX 11487575) **49**, **141**, 144
Dolaucothi, gold mine 21-22
Dowgas, tinwork (SW 964513) 32, 35
Down Tor, tinwork (SX 581696) **86**, 92
Drinick, tinbound (SW 964549) 37
Ennis, St Enoder, mineral house (SW 8449) 49
Erme Valley, tinworks (SX 639650) 22, 72, 140, **141**
Eylesbarrow, mine (SX 601683) 101
    shoad works (SX 608686) 79
Falmouth, ingot from 138
Fatt Work bounds on Dowgas Common, tinbound (SW 9651) 35
Fish Lake Foot, stamping mill (SX 64906834) **114**
Fowey Harbour, ingots from 138
Fowey, town 61

Gallidnowe New Streame in Penwith and Kerrier, tinbound 62
Gaul 21
Gobbett, blowing house (SX 64537280) 128, 135, **136**, **141**, 145, **colour plate 23**
    crazing mill (SX 64537280) 121, **136**, **141**, **colour plate 23**
Godolphin, blowing house (SW 603320) 133
    crazing stone (SW 60095188) 121
    lode-back tinwork (SW 595317) 86
Goldherring (SW 411282) 19
Good Fortune, bound at Colliford (SX 1771) 35
    tinwork on Trevedda Hill (SX 1569) 43
Good Luck, Perranzabuloe, tinbound 35
Goodfortune al's The Streamwork in Lanivet (SX 0364) 62
Goodfortune Bounds (SX 1771) 34
Goonbarne, tinbound (SW 957547) 37
Goonzion, lode-back tinwork (SX 177676) **78**, 81, **141**
    openwork (SX 177677) **78**, 92, **141**
    prospecting trenches (SX 177675) 29, 31, **78**, **141**
    shoad work (SX 176675) **78**, 79, **141**, 142-3
    tinners' building (SX 17906758) **49**, **78**
Goss Moor, finds from (SW 9459) 15, 23
    tinwork (SW 9460) 39
Goverseth, tinbound (SW 9655) 37
Great Links Tor, eluvial streamwork (SX 551862) 78, **colour plate 14**
Greenburrow, shoad work (SW 433345) 79
Greyiscome, streamwork 61-2
Hard-come-by, tinbound 37
Hardhead, lode-back tinwork (SX 150714) 81
Harrowbridge, East, eluvial streamwork (SX 200730) 76, **77**
    West, eluvial streamwork (SX 197736) **76**
Hart Tor, openworks (SX 577719) **86**, **141**, 143
    prospecting trenches (SX 577717) 29, **30**, **86**, 92, **141**, **colour plate 1**
    shoad work (SX 584724) 79
Hart Tor Brook, streamwork (SX 582717) **66**, **68**, **70**, 71, **86**, **141**, 141
    tinners' building (SX 58237169) **49**, **66**, **86**, **141**
Harvenna, stamping mill (SW 8958) 108-10, 113
Hellanoon tinwork, Lelant (SW5037) 52
Helston **11**, 152
High Moor, eluvial streamwork (SX 168797) 79
Higher Work Park, tinwork (SW 8958) 36, 43
Hingston Down (SX 3871) 29
Hingston, lode-back tinwork (SX 583691) **141**, 143
Hobb's Hill, lode-back tinwork (SX 183694) 83, **85**

openwork (SX 185714) 91
prospecting pits (SX 183694) 27
Holne Moor (SX 677711) 81, **141**, 143
Hook Lake, stamping mill (SX 63936509) **114**
Iktis 21
Ivy Tor Water (or Ladybrook), stamping mill
(SX 62859175) **114**, 128, **141**, 142
eluvial streamwork (SX 628916) 76
Jamaica Inn, finds from (SX 1877) 15
Keaglesborough, openwork (SX 572700) **86**,
92
Kenidjack Castle, enclosure (SW 355326) 19
Kerla Myne (SX 2470) 62
Kerrowe, lode-back tinwork (SW 458368) 86
Kild Nest formerly called Litle Carburlye,
tinwork (SX 1667) 43
Kit Hill, prospecting trenches (SX 377713) 29,
79
shoad work (SX 375713) **79**, 143
Kit Tin Mine (SX 565675) 99
Kregbrears, tinbounds (SW 7443) 34
Lade Hill Brook, tinners' building (SX
63928145) **141**, 144, **colour plate 4**
Langcombe Brook, alluvial streamwork (SX
602670) **71**, 72, **141**, 141
Langcombe, buddles at (SX 60376723) 128
Lanherne, finds from (SW 8765) 15
Lanlivery, finds from (SX 0759) 15
Lansdown 22
Lanyon, tinwork at (SW 4234) 36, 83
Le Halzebet in Blackmore, tinwork 39
Left Lake, stamping mill (SX 64006337) **114**
Lenobray, blowing house (SW 698473) 133-4
Leskernick, alluvial streamwork (SX 182792)
70
Leskernick, tinners' building near (SX
17887993) **colour plate 2**
Letter Moor, leats and reservoirs (SX 173702)
47, **48**, 75, 143
Liskeard **11**, 134, 152
Little Horrabridge **Colour plate 19**
Little Stamps, Porthkellis (SW 6832) 111-3
Loe Pool Valley, crazing stone (SW 6425) 121
Lostwithiel **11**, 39, 57, 61, 129, 134, 152
Lower Hartor, stamping mill (SX 60486743)
114
Ludgvan, finds from **22**
Lydford Woods, alluvial streamwork (SX
495837) 32, 61, 66, **67-69**, 70-71, **141**,
colour plate 11
Lyttle Pentrouth, tinwork (SX 0157) 36
Magor, Roman villa at (SW 637423) 22
Marazion Marsh, Roman coins from (SW
5131) 22
Meavy Valley (SX 5771) 12, **29**, **30**, **66**, **69**, **70**,
82, **86**, **91**, **127**, **141**, 141-4, **colour plates
1, 16, 17, 20 and 29**

Mendips 21-23
Merrivale, Lower, blowing house (SX
55277535) 135, **136**, **141**, 145, **colour
plates 27 and 28**
Middle, blowing house (SX 55277624) **136**,
141, 145
Upper, blowing house (SX 55187665) 116,
128, 135, **136**, **141**, 145, **colour plate 21**
Mesopotamia 20
Minear Downs, openwork (SX 032540) 91
Minions, eluvial streamwork (SX 256718) **84**,
147
lode-back tinwork (SX 259716) 81, **141**, **147**
Minzies Down, alluvial streamwork (SX
181762) 64, **65**, 66, 70
tinners' building (SX 18137604) **49**
Morvah Hill, lode-back tinwork (SW 417358)
82, 83
Mulberay, tinwork (SX0050) 39
Mulberry Hill, openwork (SX 019658) 91
Nancothan, pump from (SW 4429) 64-5
Narbo Carnon (SW 7839) 16
Newleycombe Valley, eluvial streamworks (SX
5970) 79, **86**, **141**, 142, 147
lode-back tinworks (SX 598700) 81, **86**, **141**,
143, 147
openworks (SX 593701) **86**, 91, **141**, 142,
147
Noon Digery, lode-back tinwork (SW 486352)
83
Norsworthy, Left Bank, stamping mill (SX
56786958) **114**, **141**, 145, **colour plates 16
and 20**
Right Bank, stamping mill (SX 56746954)
114
North Bounds, tinbound (SX 1771) 35
North Walla Brook, prospecting trench (SX
68068338) 29
Numphra, eluvial streamwork (SW 387292)
78
Outcombe, crazing and stamping mill (SX
58016860) **114**, 121, **141**, 145
Par Beach, ingot from (SV 932153) 22, 138
Parke an Fenten, tinwork (SW 743523) 39
Parson's Park, find from (SX 191712) 22
Paul, finds from 22
Pell, tinwork (SW 715513) 39
Penenkos, crazing and stamping mills (SW
6944) 104, 120
Penkestle, North, eluvial streamwork (SX
175703) **34**, **48**, 76, **141**
South, eluvial streamwork (SX 178699) 47,
48, 75, **85**, **141**, 142
South, lode-back tinwork (SX 178699) 83,
85, **141**, 142
Pentewan, finds from 15
Pentewan, wooden shaft (SX 0048) 15-6, 21, 63

Pentrouthe, Great, tinwork (SX 0157) 36
Penwith and Kerrier Stannary 34, 35, 37, 56-8, 62
Penwithick, ingot from (SX 0256) 20
Penzance **11**, 134, 152
Perran Arwothall (SW 7838) 44
Perran-ar-Worthal, human skeleton from (SW 7838) 15, **16**
Perranuthno, ancillary building (SW 5329) 48
Pitsloren, mine 95
Plym Valley, alluvial streamworks (SX 581660) 70
Plympton **11**, 152
Poldice, mine (SX 0050) 40, 52
   stamping mill (SX 0050) 108
Polgooth, mine (SW 9950) 35, 39
Porth streamworks, finds from (SX 0853) 15
Porth, tin ground (SX 0853) 16
Porthkellis Wartha, sett (SW 6832) 124
Porthkellis, sett (SW 6832) 38
Porthmeor, ingot from (SW 434370) 138
Praa Sands, ingot from (SW 5828) 20
Raddick Hill, eluvial streamwork (SX 579706) 76, **86**
Redhill Downs, eluvial streamwork (SX 163715) **91**, 92
   openwork (SX 163715) **91**, 92
Redhill Marsh, tinners' building (SX 16827220) **49**
Redhill, eluvial streamwork (SX 184807) 76, 79, **colour plate 13**
Relistian, accident at (SW 6036) 98
Resethdew in Kenwyn, tinbound 52
Restronguet Creek, shaft (SW 8137) 63
   tin ground (SW 8137) 16
Retallack, adit (SW 731297) 96, 145
   blowing house (SW 73122999) **116**, 130, **136**, 137, 145, **colour plate 26**
   crazing and stamping mills (SW 73182979) **114**, **116**, 121, 128, 145, **colour plates 18 and 24**
Ringleshutes, openworks (SX 674698) **141**, 143, 147
River Erme, alluvial streamwork (SX 6364) 22, 72, 140
River Taw, alluvial streamwork (SX 616904) 67
Roche, finds from 15
Rock Tin Mine, openwork (SX 014581) 91
Rosemergy, lode-back tinwork (SW 418361) 83
Rosewall, lode-back tinwork (SW 494391) 83
Rushyford Gate, eluvial streamwork (SX 216764) 76
Russells Worke, Sancreed, tinwork 36
Sampford Wood, alluvial streamwork (SX 526716) 72
Schmidmulen 104

Sheepstor Brook, float stone (SX 574675) 136
Shropshire, mines 21
Smalescombe, streamwork (SX 2074) 61-2
South Hill Leat (SX 6585) 47
South Phoenix, lode-back pit tinwork (SX 262716) **84**
South Trekeve Myne (SX 2269) 62
Spain 14
St Agnes (SW 7250) 19, 35, 39, 53-5, 103, 111, 121
   crazing stone (SW 7250) 121
St Austell (SX 0152) 35, 37, 39, 40, 44, 53, 108, 113, 129, 131
St Columb, finds from (SW 9163) 15
St Erth, finds from (SW 5535) 15
St Hilary, finds from (SW 5531) 15
St Lukes, eluvial streamwork (SX 191759) 47, 75, 79
St Margarets, tinwork (SW 9950) 39, 48
St Margitts, tinbound (SW 9950) 37
St Mawes, ingot from (SW 8433) 138
St Mewan, tin ingot from (SW 9951) 20
St Neot (SX 1867) 30-1, 34, 35, 47, 75, 132
   blowing house (SX 184679) 132
St Wenn, ingot from (SW 9664) 20
Stanlake, alluvial streamwork (SX 573714) **18**, 61, 66, **69**, 70-2
   eluvial streamwork (SX 568712) **18**, 47, 79, **91**, 92
   openwork (SX 567713) **91**, 92
Stannon Down, prehistoric settlement (SX 132803) 17
Stenaguin, tinbound (SW 9655) 37
Stertmore, tinwork (SW 601573) 40
Stonetor Brook, cache (SX647858) **colour plate 5**
Stony Girt, alluvial streamwork (SX 655663) 72
Tavistock **11**, 152
Taw, blowing house (SX 62059197) 135, **136**, 141, 145-6
Teigncombe, tinwork (SX 671865) 47
Teignhead, blowing house (SX 63778426) 135, **136**
Tinkers Bound, St Hilary, tinbound 35
Tinners' Lane, lode (SX 183697) 31
Tolcarne, Gwennap, blowing house 133
Towednack, finds from (SW 4838) 22
Trebinnick, mine (SX 183704) **96**, 101, **141**, 144
   tinners' building (SX 18367049) **49**, **96**, **141**, 144
   tinners' building (SX 18447051) **49**, **96**, **141**, 144
Treeures (Trerice), streamwork 32, 72-3
Tregenna, bank (SW 5140) 19
Treloy, finds from (SW 8562) 15, 22, **23**

Tremenhere, windmill (SW 679297) 97
Tremorwode, streamwork 61
Trerank, smelting site (SW 982598) 19
Trethurgy Round, ingot from (SX 035556) 22, 138
Trevaunance Valley, dressing floors (SW 7251) 124-5
Trevisker, prehistoric settlement (SW 8769) 19
Trewen, mine (SW 7144) 95
Trewey, lode-back tinwork (SW 457378) 83
Trewhiddle, finds from (SX 0150) 23
Trewolvas, St Columb, sett 36
Trewoone, tinbound (SW 9853) 37, 40
Trewortha, medieval farmstead (SX 226761) 55
Treyew, blowing house (SW 8144) 134
Trezelland, eluvial streamwork (SX 188793) 76
Truro **11**, 134, 152, 153
Twelve Heads, mineral house (SW 7642) 49
Tye, The, tinwork (SW 964549) 39
Tywarnhail **11**, 35, 56, **57**
Vellin Antron, blowing house (SW 7633) 133
Vitifer, mine (SX 688808) **87**, **141**, 143-4
Wedlake, eluvial streamwork (SX 536770) 76
Week Ford, Lower, blowing house (SX 66197234) 12, **136**
  Upper, stamping mill (SX 66187232) **114**
West Moor, alluvial streamwork (SX 193807) 72, **141**, **148**
  eluvial streamwork (SX 185805) **75**, 76, **141**, **148**
West Park, tinwork (SX 0454) 37
Westmoorgate, eluvial streamwork (SX 201804) 76, **77**
Wheal Prosper, openwork (SX 028642) 92
Wheal-an-Venton, Gwennap, tinbound 37
Wheale an Gours, St Agnes, tinbound 35
Wheale Bargus, mine (SW 5931) 96
Wheale Bregge, mine (SW 5930) 96
Whele an Fenten, tinwork (SW 5038) 39
Whele an Prat, tinwork (SW 717516) 39
White Worke Bounds, tinbound (SW 5431) 35
Witheybrook, prehistoric settlement (SX 244742) 17
Yealm Valley, prospecting trench (SX 61806360) 29
Yealm, Lower, blowing house (SX 61786351) 12, 128, **136**, **141**, 146, **146**
  Upper, blowing house (SX 61716385) 135, **136**, **137**, **141**, 146, **146**
Yellowmead, crazing mill (SX 57426755) 121

General Index

Abraham the Tinner 40, 61
accidents 46, 98
accommodation 50-5
adits 22, **29**, 35, 42-3, **79**, 86, 90, 93-95, **96**, 97-98, 101-2, 143-4, 148, 149-51
Agricola, G. 26, **27**, 31, 44, 94-9, **100**, 101-2, 104-5, **106**, 120, **122**
Alcock, L. 23
alder 16-7
ancillary buildings 12, 40, 47-50, **66**, **85**, **86**, 96, 140, 143-5, 149 **colour plates 2, 3 and 5**
Anonymous writer, The 26-9, 31, 50, 95, 98, 102, 104, 121-4, 132, 134
archaeological excavations 12, 17, 19-20, 22, 49-50, 55, **68**, **69**, 71, **90**, 107, 116, **117-9**, 120, 123, 125, 127-8, 135, 145, **colour plates 6, 11, 21, 22 and 25**
artefacts 12, 14-5, 23, 50, 72
Austin, D. **90**, **117**, **119**, **126**
Beagrie, N. 22, 138
beams (see openworks)
Beare, Thomas 40-1, 48, 62, 73, 129-31, 134, 137
bellows 20, 97, 129, **130**, 133, 149
Black Prince 32, 62, 72-3
black tin 20, 36, 49-50, 62, 104, 110, 122, 124, 129, 132, 134, 149
Blanchard, I. 50, 52
blowing house, marks 130, **131**
blowing houses 12, 104, **116**, 122, 129, **130**, 131-134, **135**, **136**, 137-8, 145-6, **146**, 149, 152, **colour plates 21, 23, 27 and 28**
Borlase, W. 31, 73-4, 125
bounding 32-8, 48, 118, 134
bounds 32-40, 43, 45, 48, 52, 56, 62, 87, 96, 108, 118, 120, 134, 152
bowl furnaces 14, 19, **20**, 129
boys 40, 111-2, 123-5
bronze, 14-6, 22, 139
Bronze Age 11, 14-21
Brown, A.P. 16-7
Bryant, J. 12
buddles 66, 71, **105**, **107**, 108, 111, 117, **118**, 121, **122**, 123-5, **126**, 127-130, 142, 145, **146**, 149, **colour plate 25**
Burnard, R. 12
caches **49**, 50, 149, **colour plates 5 and 6**
capitalism 11, 25-6, 32, 37, 38-40, 42-4, 113-4
Carew, Richard 15-6, 26, 28, 30, 36, 38, 41, 43, 95, 97-8, 102-6, 123-4, 130, 132-4
cassiterite 14-5, 17, 19, 22-3, 25, 32, 38, 40, 42, 44, 47, 50, 54-5, 60-2, 66, 72, 80-1, 93,

104, 107, 112, 117, 121, 123, 125, 134, 142, 149, 152

cessation 44-6, 114

charcoal 11, 16, 20, 98, 129-32, 149

clash mills (see stamping mills)

coffins (see openworks)

coinage 57, 110, 113, 124, 132, 134, 136, 152

coinage town 52, 134-5, 138, 152

collapse 46, 97-8

Collins, J.H. 31, 64-5, 70

Cornwall Archaeological Unit (formerly C.C.R.A.) 19, **28**, **74**, **75**, **77**, **79**, 81, **82**, **84**, 86, **147**, **148**

costeening 26, 30

crazing mills 104-5, **116**, 119, 120-2, 145, 149, 152, **colour plates 23 and 24**

cuesta shaped earthworks 64, **65**, 67, 69

Dark Ages 23-4

Dartmoor Tinworking Research Group 12, 142, 145

dating 10, 17, 22-3, 58, 62, 65, 71, 81, 92-3, 101, 104, 117, 119-20, 129, 137-8

deforestation 98, 130-1

Diodorus Siculus 21

dowsing **27**, 31-2

drainage 43, 45, 66-7, 93, 95-7, 149-50

dreams 30-1

dressing 42, 47, 66, 71, 79, 80, 104, **105**, **107**, 108, 111, 114, 117, **118**, 120, 121, **122**, 123-5, **126**, 127-30, 142, 145, **146**, 147, 149, 152, **colour plate 25**

dry stamping 105-6, 119-20

Earl, B. **94**

environmental evidence 14, 16-7, 19, 22-3, 65, 118

essay Hatches 26-7

exhaustion 22, 35, 44-6, 58, 59, 63, 90, 93-4, 120, 139

feldspar 60

Fiennes, Celia 40

Finberg, H.P.R. 24

fire-setting 91-92, 103

Fleming, A. 17

float stones 129, **130**, 136, 145, 149, 150, **colour plate 28**

flooding 46, 61

Foweymore Stannary **11**, 12-4, 16, **18**, 36, 37, **51**, 52, 55, 56, **57**, **59**, 61, 62, 81, **83**, 87, **88**, 107, **115**, 132, 134

furnaces 11, 19-20, 23, 129-36, 145, **146**, 149, **colour plates 21, 27 and 28**

gentry 38-9, 52

Godley, A.D. 20-1

granite 60, 129, 134-6, 149-50

Greeves, T. 12, 124, 132, 136, 138

Griffith, F. **87**

gunpowder 11, 91, 103

Hamilton Jenkin, A.K. 15, 43, 81, 131

Hartgroves, S. **28**, **74**, **75**, **77**, **79**, **82**, **84**, **147**, **148**

Hatcher, J. 23, 25, 35, 42, 46, 50, 52, 58, 73, 130

hatches 26-27, 32, **33**, 65-7, **68**, 112, 150

hatchworks 62-3, 65-7

health 101-2

Henderson, C. 153

Henwood, W.J. **16**, **23**, 31

Herring, P. 86

Hitchens, F. and Drew, S. 63-5, 72

Hunt, R. 22

ingots 14, 20, 22, 25, 129-30, 134, 136-8, 150

Iron Age 15, 19-20

Jews 15

Kelly, J. 12, 145

kieves **107**, 125

knocking/knacking mills (see stamping mills)

ladders **94**, **100**, 101-2

leats 28-30, 34, 47, **48**, 61, 65-6, **68**, 73, **74**, 75, 83, **85**, **86**, **89**, 92, 108, 110-2, **117**, **118**, 119-20, 125, 140, 143, 145-6, 150

Leifchild, J.R. 31

Lethbridge, D. **colour plates 5, 16, 17, 19, 23 and 28**

level **34**, 63-4, 66-7, 90, 93, **94**, 95-6

Lewis, G.R 15, 24, 25, 45, 62

lode-back pit tinworks 27-8, **29**, 45, 79, 81, **82**, **83**, **84**, 85, **86**, 87, 92-3, 96, 107, 143, **147**, 150

lodes 15, 19-20, 25, 27-32, 40, 42-3, 58, 60, 62, 72-3, 81, 86-7, 89-90, 92-6, 101-3, 107, 114, 123, 132, 134, 143, 149, 150-1, 152

Maclean, J. **59**

medieval settlements **51**, 54-5, 142, **colour plate 7**

melior stones **99**, 101, 144, 150

merchants 35, 38-9, 42, 45, 52, 113, 133, 138

mica 60

mining 12, **29**, 38, 40-6, 58-9, 62, **78**, **79**, 81, **82-91**, 92, 93, **94**, 95, **96**, 97, 98, **99**, **100**, 101-3, 143-4, **147**, 150

mortar stones 105-6, **115**, 116-8, 144-5, 150, **colour plates 16-20 and 24**

mould stones 129, **130**, 135-6, **137**, 138, 145, **146**, 150, **colour plates 19, 21, 26, 27 and 28**

Muhly, J.D 20-1

Newman, P. 12, **105**, **130**

Norden, J. 22, 26, 47-8, 62, 98

Northover, J.P. 14-5

old men 81

openworks 12, 45, **78**, **79**, 81, **86**, **87**, **88**, **89**, **90**, **91**, 93-4, **99**, 107, **117**, 119-120, 142, 143-4, 147, 150-1, **colour plate 15**

parallelworks **66**, **68-70**, 71-2, **colour plate 9**
peat drying platforms **75**, **77**, 132
Penhallurick, R. 16
Pennington, R.R. 25, 32-3, 36-7
Powell, C. **colour plate 8**
prehistoric settlements 17, **18**, 19-20, 49, 143
Price, D. 17
prospecting 25-6, **27-30**, 31-2, **34**, 46, 66, **78**,
    85, 92, **96**, 142-4, 150, **colour plates 1**
    and 8
Pryce, W. 26, 31, 44, 64-5, 70, 87, 89-90, 92,
    104, 106, **107**, 110, 130
pumping 11, 45, 64-5, 83, 96-7
quartz 60
raising ore 98-101
reservoirs **28**, 29, **30**, **34**, 47, **48**, **74**, 75, **76**, **86**,
    89, **91**, 92, **116**, 142, **colour plate 8**
retained dumps 65, **71**, 72, 140, **colour plates**
    12 and 14
reverbatory furnace 11, 131
reworking 42-4
Roman period 21-3, 138
Royal Commission on the Historical
    Monuments of England [RCHME] 12,
    81, 87
sample moulds 138
Scrivener, R. 123
sett 36-8, 46, 54, 112-3, 124
shafts 12, 21, 63, 79, 81, 87, 92-3, **94**, **96**, 97-8,
    99, 101-2, **117**, 143, 144, 147
Shell, C.A. 16
shelters 12, 47-50, **66**, **85**, **86**, 96, 140, 144-6,
    colour plates 2, 3 and 5
shoad 26-7, 29, 31, 72, 79-80, 143, 150
shoad works **78**, **79**, 80
shoring mines 97-8
Simmons, I.G. 16
slag 19, 115, 129, 132, 134-5, 145
smelting 12, 14, 16, 19, **20**, 104, 115, **116**, 122,
    125, 129, **130**, 131-4, **135**, **136**, 137-9, 145,
    146, 149, 152, **colour plates 21, 23, 27**
    and 28
smuggling 130, 137-8
spalliard 40-1
stamping machinery 11-2, 104, **105**, **106**, 107,
stamping mills **105-8**, 109-13, **114**, 115, **116-9**,
    120, **127**, **colour plates 21, 22, 23, 24 and**
    29
stamps (see stamping machinery)
Stannaries 11, 24, 32-3, 35, 38, 48, 52, 54-9,
    62, 137-8, 152
Stannary Court Rolls 32-3, 36-9, 46, 48, 56,
    59, 62, 152
    Law 25, 35
    town **11**, 152
steam engines 11
stiling 63-4, **71**, 72, **colour plates 12 and 14**

stopes 91-3, **94**, 102, 152
storage buildings **49**, 50, **colour plates 5 and 6**
streamworks, alluvial 33, **51**, 60-4, **65-71**, 72,
    73, **84**, **86**, **117**, **127**, **148**, **colour plates 9,**
    11 and 12
    eluvial **34**, **51**, 72, **74-7**, 78-9, **85-7**, **91**, **147**,
    148, **colour plates 8, 10, 13 and 14**
    tailings 63-4, 71, 75, 78, 92, 112, 120-1, 123,
    125, 127-8, 149, 150
tailrace **105**, 145
Threipland, L.M. 19
tin ground 15-6, 22, 60-1, 63-4
tin mills 12, 30, 35, 44, 46-7, 96, 104-21, 129-
    35, 140, 143, 144-5, 149-52
toll Tin 35-7, 52, 56
tools 50, 102-3, 149
trayning 26-7, 32
tribulage 57
truck system 41
tyes 39, 60-1, 63-5, 70-2, 76, 78, 122, 125, 152,
    colour plate 8
Tylecote, R.F. 14, **20**
ventilation 91, 95, 97
Wacher, J. 22
washes 124
waterwheels 96-7, **105**, 106, **107**, 119-20, 129,
    130, 144, 149, 152
westward production shift 57, 58, **59**
wet stamping **105**, **106**, 107, 119-21
wheelpit 83, 106, 115, 117, 120, 144, 152
Whetter, J. 113, 130-1
whim platforms **94**, **96**, 99, 101, 143, 144, **147**,
    150
windlass **94**, 98-9, **100**, 101
winze **94**
women 39, 44, 125
work force 40-2, 112
Worth, R.H. 12, 41